Pathogens Crossing Borders

The increasing globalisation of trade, travel and transport since the mid-19th century had unwelcome consequences—one of them was the spread of contagious animal diseases over greater distances in a shorter time than ever before. Borders and national control strategies proved to be insufficient to stop the pathogens. Not surprisingly, the issue of epizootics (epidemics of animals) was among the first topics to be addressed by international meetings from the 1860s onwards. *Pathogens Crossing Borders* explores the history of international efforts to contain and prevent the spread of animal diseases from the early 1860s to the years after the Second World War. As an innovative contribution to global history and the history of internationalism, the book investigates how disease experts, politicians and state authorities developed concepts, practices and institutional structures at the international level to tackle the spread of animal diseases across borders. By following their activities in dealing with a problem area which was—and is today—of enormous political, social, public health and economic relevance, the book reveals the historical challenges of finding common international responses to complex and pressing global issues for which there are no easy solutions.

Cornelia Knab is Head of Scientific Management and Research Associate at the Europainstitut / Institute for European Global Studies of the University of Basel, Switzerland.

Routledge Studies in the History of Science, Technology and Medicine

Soviet Science and Engineering in the Shadow of the Cold War
Edited by Hiroshi Ichikawa

Cold Science
Environmental Knowledge in the North American Arctic during the Cold War
Edited by Stephen Bocking and Daniel Heidt

Medical Memories and Experiences in Postwar East Germany
Treatments of the Past
Edited by Markus Wahl

Politics, Statistics and Weather Forecasting, 1840–1910
Taming the Weather
Edited by Aitor Anduaga

Social Class and Mental Illness in Northern Europe
Edited by Petteri Pietikäinen and Jesper Vaczy Kragh

Medicine and Justice
Medico-Legal Practice in England and Wales, 1700–1914
Edited by Katherine D. Watson

Lifestyle and Medicine in the Enlightenment
The Six Non-Naturals in the Long Eighteenth Century
Edited by James Kennaway and H.G. Knoeff

Germs in the English Workplace, c. 1880–1945
Edited by Laura Newman

Pathogens Crossing Borders
Global Animal Diseases and International Responses, 1860–1947
Edited by Cornelia Knab

For more information about this series, please visit: https://www.routledge.com/Routledge-Studies-in-the-History-of-Science-Technology-and-Medicine/book-series/HISTSCI

Pathogens Crossing Borders

Global Animal Diseases and International Responses, 1860–1947

Cornelia Knab

Routledge
Taylor & Francis Group
LONDON AND NEW YORK

First published 2022
by Routledge
4 Park Square, Milton Park, Abingdon, Oxon OX14 4RN

and by Routledge
605 Third Avenue, New York, NY 10158

Routledge is an imprint of the Taylor & Francis Group, an informa business

British Library Cataloguing-in-Publication Data
A catalogue record for this book is available from the British Library

Library of Congress Cataloging-in-Publication Data
Names: Knab, Cornelia, 1980- author.
Title: Pathogens crossing borders : global animal diseases and international responses, 1860-1947 / Cornelia Knab.
Description: Abingdon, Oxon ; New York, NY : Routledge, 2022. | Series: Routledge studies in the history of science, technology and medicine ; 45 | Includes bibliographical references and index.
Identifiers: LCCN 2021052985 (print) | LCCN 2021052986 (ebook) | ISBN 9780367471880 (hardback) | ISBN 9781032251868 (paperback) | ISBN 9781003033998 (ebook) | ISBN 9781000572643 (adobe pdf) | ISBN 9781000572667 (epub)
Subjects: LCSH: Veterinary epidemiology--History. | Communicable diseases in animals--Prevention--International cooperation. | Communicable diseases in animals--Prevention--Government policy. | Animal health--International cooperation. | Animal health--Government policy.
Classification: LCC SF780.9 .K53 2022 (print) | LCC SF780.9 (ebook) | DDC 636.089/44--dc23/eng/20211207
LC record available at https://lccn.loc.gov/2021052985
LC ebook record available at https://lccn.loc.gov/2021052986

ISBN: 978-0-367-47188-0 (hbk)
ISBN: 978-1-032-25186-8 (pbk)
ISBN: 978-1-003-03399-8 (ebk)

DOI: 10.4324/9781003033998

Typeset in Sabon
by SPi Technologies India Pvt Ltd (Straive)

Contents

Acknowledgements

On the long way from the choice of topic to the finished book, numerous people have contributed in different ways to the realisation of this project. Without their interest, their enthusiasm to discuss my ideas and their great help, this book would not have been possible. I am deeply grateful to all those who have encouraged, inspired and supported me in my work.

This book is the revised and abridged version of my doctoral dissertation submitted in 2013 to the Faculty of Philosophy at Heidelberg University, originally entitled "'Organising Trust': International Cooperation against Animal Diseases, 1863-1946". I would first of all like to profoundly thank my supervisor Madeleine Herren. My curiosity about the history of animal diseases was awakened by her in the last year of my studies at Heidelberg University. From the very beginning she encouraged my interest in this subject and involved me intensively in diverse research projects. Her motivation to 'think outside the box' has shaped my approach to global history, and her thought-provoking impulses and challenging questions have contributed significantly to the development of this book. I would further like to express my profound gratitude to my second supervisor, Roland Wenzlhuemer, for his dedicated support at all times, his constructive advice, and his valuable impulses, especially on the subject area of global history.

The research on the book took me to many archives and libraries in Europe and the US. I would like to thank the staff of numerous institutions for their assistance and friendly advice from which I benefitted a lot. Most notably, the archivists at the French Archives Diplomatiques (La Courneuve), the League of Nations Archives and the ILO Archives (Geneva), the Schweizerisches Bundesarchiv (Berne), and the Politisches Archiv des Auswärtigen Amts (Berlin) have provided valuable and useful support. I am especially grateful to Marie Teissier (OIE, Paris) for her great help concerning the history and documents of the OIE. I would also like to express my thanks to the librarians of the Veterinärmedizinische Bibliothek of the Freie Universität Berlin, the library of the Stiftung Tierärztliche Hochschule Hannover, the Bibliothèque interuniversitaire de santé Paris, the libraries of the Universities of Heidelberg and Basel, and the Schweizerische Vereinigung für Geschichte der Veterinärmedizin, as well as to the staff of the Musée

Fragonard (Maisons-Alfort) for their inspiring explanations of veterinary history.

The book was completed during the COVID-19 pandemic, providing new aspects to the topic area in its global dimension. While I worked on the book, many people, most of them in Germany, the US, the UK, Switzerland and France, have supported me. When I started this project, I was integrated in a research project at the Cluster of Excellence "Asia and Europe in a Global Context" at Heidelberg University. I benefitted enormously from the interdisciplinary research environment, the lively exchange, and the transcultural encounters which have led to unforgettable memories. Many friends and colleagues at the Cluster of Excellence, the Department of History of Heidelberg University and the Europainstitut/Institute for European Global Studies of the University of Basel have supported, encouraged, and inspired me, or helped me with their critical thoughts and comments at different stages of the projects. My profound thanks go to (in alphabetical order) Benjamin Auberer, Caroline Authaler, Davina Benkert, Dominique Biehl, Eric Decker, Amanda Dixon, Isabelle Dörr, Anne Gardner, Matthew Gardner, Frank Grüner, Milena Guthörl, Laura Jakobs, Lars Kury, Toshiki Mogami, Sergio Mukherjee, Annette Must, Haifen Nan, Maya Okuda-Yazaki, Corinne A. Pernet, Ines Prodöhl, Amalia Ribi Forclaz, Ina Serif, Christiane Sibille, and all those not named here who have supported me. I also would like to thank Max Novick and the Routledge team for their patience and support.

I would like to profoundly thank Christian Futter for his amazing support and great encouragement. My special and heartfelt thanks go to the Knab family—Herbert, Gisela, Christoph and Nina Knab, and all other family members—for their constant support, their great sense of humour and their encouragement to keep up scientific curiosity. This book is dedicated to them.

Abbreviations and acronyms

AIDS	Acquired Immune Deficiency Syndrome
ADLC	Archives Diplomatiques, France, La Courneuve
BAI	Bureau of Animal Industry (United States Department of Agriculture)
BAR	Schweizerisches Bundesarchiv/Swiss Federal Archives
BSE	Bovine Spongiform Encephalopathy
BTW	Berliner Tierärztliche Wochenschrift
COVID-19	Coronavirus Disease of 2019
DKZ	Deutsche Kongress-Zentrale
FAO	Food and Agriculture Organization of the United Nations
GARC	Global Alliance on Rabies Control
GATT	General Agreement on Tariffs and Trade
IIA	International Institute of Agriculture/Institut International d'Agriculture
IIR	International Institute of Refrigeration
ILO	International Labour Organization
LN/SDN	League of Nations/Société des Nations
LON	League of Nations Archives
LNHO	League of Nations Health Organisation
OEEC	Organisation for Economic Cooperation in Europe
OIE	Office International d'Epizooties/World Organisation for Animal Health/Organisation Mondiale de la Santé Animale
OIHP	Office International d'Hygiène Publique/International Office of Public Hygiene
PAAA	Politisches Archiv des Auswärtigen Amts, Berlin, Germany
PAHO	Pan American Health Organization
PANAFTOSA	Pan American Foot-and-Mouth Disease Center
SS	Schutzstaffel
SARS	Severe Acute Respiratory Syndrome
UN	United Nations
UNICEF	United Nations International Children's Emergency Fund
UNRRA	United Nations Relief and Rehabilitation Administration
U.S.	United States of America
WHO	World Health Organization

1 Introduction

The global spread of animal diseases and international responses

In the summer of 1920, the news of an outbreak of rinderpest in Belgium raised concerns among governments and veterinary experts in Europe. Zebus from India, to be transported to Brazil via the port of Antwerp, had transmitted the highly contagious viral disease to livestock in Belgian stables.[1] It turned out that the appearance of the cattle disease in Belgium was not an isolated case in Europe during those years. As a consequence of warfare with its movements of troops and trails of refugees, accompanied by livestock, outbreaks of rinderpest occurred from 1918 to the early 1920s in several countries of Eastern Europe, as well as briefly in Greece and in Italy.[2] The European governments had good reasons to be concerned about rinderpest and the animal disease situation in general. For centuries, epizootics (epidemics of animals) had on several continents severely damaged the economic livelihood of societies depending on animal husbandry. Rinderpest could rapidly spread among different species of domestic and wild animals with an extremely high rate of mortality, and outbreaks often resulted in devastating consequences for livestock owners. Moreover, the news about rinderpest in Europe came at a very difficult time for the war-stricken livestock sector—several European countries were facing severe outbreaks of foot-and-mouth disease in the same year.[3] Another highly contagious viral animal disease, foot-and-mouth disease occurred repeatedly in many regions of the world. It was—and still is—one of the most dreaded animal diseases due to its easy transmission among animals and the massive economic damage it can cause to livestock farming.

In response to the outbreaks of foot-and-mouth disease and rinderpest in Europe, the French government took the initiative of organising an international conference to address the problem of the cross-border spread of animal diseases. As a result, in May 1921 the Conférence Internationale pour l'Etude des Epizooties (International Conference for the Study of Epizootics) convened in Paris to discuss the current epizootics situation and future international cooperation in the prevention and control of contagious animal diseases.[4] The participants of the international meeting had to deal with

DOI: 10.4324/9781003033998-1

questions that have today lost nothing of their urgency: which strategies and measures can effectively control highly contagious animal diseases and prevent their spread in an increasingly interconnected, globalised world? How can international cooperation be organised to tackle the problem of animal diseases cutting across borders in an effective way—given a situation of politically and ideologically often antagonistic states? The scientists, politicians and diplomats who met in Paris in 1921 were aware of the complexity and global interconnections of animal disease problems and their potential impact on economic prosperity and public health. They agreed that existing strategies and measures conducted at the national level were insufficient given the spread of rinderpest, foot-and-mouth disease and other contagious animal diseases, and that a new international initiative to address the problem area, especially with regard to its transboundary dimension, was urgently needed. They opted for an established approach of internationalism—the creation of a new international office. In January 1924, an international convention confirmed the foundation of the Office International des Epizooties (OIE) in Paris.[5] Three years later, in March 1927, the OIE began its work with the first session of its international committee. And 90 years after the conference at Paris, in May 2011, the OIE—since 2003 called World Organisation for Animal Health—could celebrate a milestone in the history of epizootics control. Together with the Food and Agriculture Organization of the United Nations (FAO) it declared rinderpest to be the first animal disease to be eradicated on a global scale.[6] The global end of rinderpest in 2011 was a spectacular success of collaborative international efforts in the area of animal diseases: it was the result of a long-term and multi-levelled cooperation between a network of international organisations, national veterinary administrations, research institutions, and multiple regional and local authorities and partners all over the world.[7]

Rinderpest was only one of many animal disease problems that materialised as threats to animal and human health, as well as to economic prosperity, at both global and local levels. This book explores the history of international efforts addressing the spread of animal diseases in the late 19th and first half of the 20th century. It investigates how scientific and medical experts, politicians, diplomats and international officials developed concepts, practices and institutional structures to tackle animal diseases at the international level to stop and prevent their cross-border spread. By examining their responses to problems of animal diseases, the book aims to provide insights into the history of the development of international strategies on pressing global issues of entwined economic, political and public health significance. The emergence, re-emergence and spread of animal diseases and their potentially far-reaching consequences also for human health—in particular the threat of new pandemics—are today regarded as major global problems of the present and future, especially as new globally relevant diseases appear every year. Since the emergence of the COVID-19 pandemic, the issue in its global and animal-related context has moved strongly up the political agenda and into the public consciousness. What can a historical approach contribute to such a subject area known for its present-day global significance? A historical perspective on international efforts addressing animal diseases, this study underlines, does not offer solutions to today's global animal disease problems. Examining historical approaches towards animal disease problems, the book suggests, uncovers challenges that arise when societies seek answers to complex cross-border problems in a globally intertwined and interdependent world. It provides a sharpened focus and differentiated insights on how historical actors in specific historical constellations perceived problem areas in their interconnectedness, and how they problematised their global dimension. The historical perspective focuses on investigating the development of strategies and concepts to address globally relevant problems in the absence of clear solutions, and on the processes of creating international norms and standards in their social, economic and political contexts. It provides a lens into historical power relations, prioritisations and the enforcements of agendas, and the inclusion or exclusion of historical actors and their agencies in international approaches towards a global problem area. Thus it helps to critically examine the historical genealogies of international cooperation in efforts of disease control, and their long-term effects till today. It shows that neither is there a linear story of progress nor are there simple answers or solutions when addressing a complex, multi-levelled problem area such as contagious animal diseases and their spread. A historical perspective on this topic area, this book suggests, contextualises and problematises narratives of international approaches, and contributes to a differentiated and better understanding of the ambivalent historicities of globally oriented activities to control and contain contagious diseases.

The living environments of humans and animals were in the past, and are at present, closely intertwined on many levels. Diseases that can be transmitted between different species, so-called zoonoses, not only constitute economic problems for the livestock sector, but can also, in the case they are transmitted to humans, directly endanger human health. The OIE estimates that about 60 percent of today's infectious human diseases are of animal origin, including Ebola, HIV, influenza, and SARS, as well as COVID-19 caused by a coronavirus.[8] Serious outbreaks of epizootics, and of human epidemics of zoonotic origin, have been a recurrent feature in history. Human activities such as livestock husbandry, hunting or pet-keeping contributed to the emergence of contagious diseases and facilitated their spread, which was further enhanced by activities of travel, trade, colonialism and warfare. In all historical periods, the outbreaks of severe epizootics had profound political, economic and social repercussions. During the 19th century, when initiatives on an international level addressing animal disease problems began, contemporaries perceived a number of transformations in the emergence and spread of animal diseases, linked to the intensification and expansion of globalised capitalism, industrialisation and colonialism. While animal diseases had always had the potential to spread transcontinentally over large areas, in the context of the increasing global interdependencies since the 19th century they could in many cases be transmitted over greater distances in a shorter time than before.[9] Especially in the second half of the 19th century, the worldwide trade volume in livestock and animal products increased significantly, as the new steamships and railways allowed the transport of previously unimaginable numbers of animals and goods. Technical innovations in packaging and in the preservation of food enabled a better trading of perishable animal products over long distances. Warfare—with the corresponding movements of people and animals—was frequently conducted on a global scale. The capitalist demand for more and more natural resources, the intensified activities of colonialism, and the growing hunger for land for agriculture and settlements contributed to the movements of large numbers of people into previously thinly populated areas of the world, which were made accessible by new roads, shipping routes and railway lines. The resulting contacts with wildlife contributed to bring new animal diseases to the fore and to spread diseases which before usually had appeared locally. At the same time, the understanding of animal diseases underwent profound changes due to new developments in human and veterinary medicine as well as in the natural sciences. The scientific identification and description of a large number of pathogens and the research on their contagiousness and their pathways of transmission opened up new approaches for the diagnosis, treatment and prevention of diseases. Scientific research also contributed to making industrial forms of factory farming more profitable, which in turn also changed the conditions for the emergence and spread of diseases.[10]

The international initiatives in the thematic area of animal diseases beginning since the 1860s were thus an attempt to tackle a transboundary problem perceived to be, at least to a large degree, a result of technical progress and increasing global interconnections. After the First World War, several international organisations integrated the thematic area of animal disease control into their agenda, among them the League of Nations and the International Labour Organization (ILO) in Geneva and, as already mentioned, the OIE in Paris. Following the efforts of the veterinary and medical experts, scientists, diplomats and civil servants active in these international contexts since the middle of the 19th century until the end of the Second World War, this book explores their discussions and interactions about strategies, interdependencies and responsibilities in dealing with a complex and globally interconnected area. They developed what is referred to in this book as veterinary internationalism—the efforts of historical actors to create professional networks, platforms of exchange, and institutional structures for cooperating at the international level to address animal disease problems, especially with view to their transboundary spread. How did historical actors develop concepts, practices and institutions to address a problem area cutting across borders in the context of competing national interests and the "mechanics of internationalism"?[11] Who was part of these international activities, who were the excluded and the silenced voices? What were the implications of hierarchies and asymmetries on the international level for the control of animal diseases in a global context? And how does the analysis of the historical international activities to contain animal diseases help us to improve our understanding of the power structures and ambivalences of international efforts in dealing with urgent problems of a global dimension?

In examining historical international approaches to animal disease problems, this book focuses in particular on two interconnected thematic strands. First, it addresses the interrelations and tensions between the European, the international and the global scales that were shaping the discussions, concepts, practices and institutional structures developed to tackle the problem of the cross-border spread of animal diseases. The debates on and interpretations of these scales at international platforms, this study underlines, had a formative influence on the activities addressing cross-border animal disease control. It was especially the dominant focus on European concerns and the related power asymmetries on both the political and the scientific level that became a characteristic, as well as gradually a main problem, of international efforts to address animal diseases. With regard to the early history of international public health, Alison Bashford has outlined the interrelation and quasi-congruence of the 'international' and 'European':

> At the beginning of the 20th century, 'international' often signified Europe: the so-called 'international' health organizations (at least the early examples like the Paris Office [the Office International d'Hygiène Publique] and the League of Nations Health Organisation) were at core, in orientation, and in interest, regional European organizations.[12]

Bashford encourages to take into account the need to "provincialize Europe" (Dipesh Chakrabarty)[13] and, accordingly, to reflect on the interconnected regional, national and colonial dimensions in their global contexts in the history of international health policies.[14] This assessment is also important with regard to the topic of international animal disease control. As this study shows, the narratives and concepts of what constituted 'Europe', 'international' and 'global' in the context of international veterinary discussions were neither unambiguous nor uncontested. Similar to the discourses in the area of epidemics and public health, international discussions and projects on animal diseases in the 19th and early 20th century focused mainly on European-related disease concerns, in particular on the task of protecting European territories from animal diseases. The priority of European agendas on international platforms discussing animal diseases, and the constructed congruence of the regional European with the international scale, was omnipresent—reflected in epistemological categories, scientific concepts, anti-disease practices and institutional developments, as well as in the locations and the list of participants of international meetings. The constructed dichotomy between 'Europe' and other areas of the world was linked to ideas of 'civilised' states and their sanitary veterinary approaches, measured in this case by the quality of the national or colonial veterinary system according to Western and Central European standards.[15] The focus on European concerns characterising the international concepts, practices and institutions of animal disease control had long-term consequences. Indeed, the Europe-centred approaches of veterinary internationalism, given the globally intertwined dimension of the problem area, proved to be increasingly insufficient to successfully and effectively tackle animal disease problems, even in the eyes of experts interested in the 'protection' of European territories. As this study underlines, the predominant focus on European concerns in international approaches gradually turned out to hamper a comprehensive approach towards animal diseases, given their globality, their cross-border spread, and their complex interdependences with other problem areas. The global scale was, simultaneously, very present in the international discussions on animal diseases: many historical actors linked the global scale of animal diseases primarily to their possible origin and movements, the associated uncertainties regarding their temporal and spatial extent, and thus highlighted the dangerous dimension of global interconnectedness and interdependencies for Europe. The problem of how to find ways to deal with these dangerous and uncertain consequences of global connections—especially those being a threat to European economies—was a central issue in international discussions. At the same time, internationalism underwent significant transformations. From the 1920s onwards, more and more representatives of non-European countries were present at international meetings dealing with the cross-border spread of animal diseases. By this time, several non-European countries had established themselves as important producers of meat and other animal products for the world market. But the focus on European agendas in international organisations and projects on animal diseases remained, and it was underlined by the fact that the majority of international meetings on topics of animal disease control took place in Western or Central Europe. In the field of international public health, governments addressed this problem by creating regional offices beyond Europe, notably the International (Pan American) Sanitary Bureau, the predecessor of today's Pan American Health Organization (PAHO). Similar non-European long-term international initiatives in the area of animal health were rarely established on the level of institutions until the mid-20th century.[16]

A second thematic focus of this study concentrates on the historical actors and their efforts to develop international strategies to address the cross-border spread of animal diseases. Since the middle of the 19th century, for some governments, the active involvement in internationalism, in particular the hosting of international organisations and conferences, was a means to increase political influence and prestige.[17] With regard to questions of animal disease control, governments were interested in the benefits of international cooperation to support their own livestock economies. But many were hesitant to participate in legally binding commitments of international cooperation beyond the level of bilateral agreements.[18] This attitude was also visible in many other areas of international cooperation. Colonies, in any case, were only involved to a limited extent in international activities against animal diseases. Veterinary experts, the main historical protagonists on the international platforms, often combined several professional positions and duties: of a high-ranking official within the national veterinary services, a scientific researcher, a member of a national and a transnationally networked scientific community, a veterinary practitioner, and a national representative in an international meeting. On international platforms, they actively used and influenced the instruments of international organisation. They tried to initiate formal international agreements in order to advance long-term governmental cooperation, but as they soon learned, international conventions were enormously difficult to negotiate, establish and implement. This book proposes to shift the focus from the dichotomous question about 'success' or 'failure' of international initiatives to consider the long-lasting impact, repercussions, lessons and side effects of the work and the commitment of historical actors in historical veterinary internationalism. The study argues that, in the case of animal disease control, the processes of discussion, conceptualisation and negotiation of international agreements—also in case they were not or not much implemented in practice—were important for the development of international strategies against contagious animal diseases. They enabled historical actors to engage in multilateral learning processes, activities of networking and long-term international and national agenda setting both within and beyond official frameworks. As this study shows, veterinary experts did not act merely as providers of technical knowledge and advice for political decision makers, but operated with their professional authority to initiate and expand cross-border strategies against diseases, despite their position of, in most cases, limited direct political power. Many showed interest not only in activities of scientific exchange, but in promoting joint projects of international cooperation to contain diseases. The study emphasises that the activities of technical-scientific, in this case veterinary, internationalism cannot be considered as inferior to diplomatic work in advancing international cooperation activities: expert internationalism, diplomatic negotiations, and semi-official, professional or private international activities were often interconnected and mutually interdependent. To explore and analytically grasp the international activities of veterinary experts, the book suggests the conceptual idea of 'organising trust'. The notion goes back to a statement by Ludwik Rajchman, director of the Health Organisation of the League of Nations and one of the important personalities of international health in the interwar period.[19] At a meeting of experts at the League of Nations in early 1928 about the question of how to facilitate the international trade with livestock, Rajchman stressed the urgency of improving the quality and transparency of national sanitary administrations in order to make effective international cooperation against both human and animal diseases possible. He emphasised the need to "organise trust"[20] between international partners in order to enhance mutual security against infectious diseases.

Based on this statement by Rajchman, I suggest using the idea of 'organising trust' in order to understand international organisations as instruments and platforms for developing and negotiating trust-building measures and activities on animal disease control. The concept of 'organising trust' helps in this study to analytically explore the international strategies, concepts, and communication efforts of historical actors. Relationships based on the development of forms of trust were essential to convince potential partners to join activities of international cooperation, to advance common scientific projects, to develop systems for transparent information transfer, to use global communication networks for the establishment of disease warning systems, and to search for alternative ways of cooperation in times of political and economic tensions.

Animal disease histories beyond borders

Until the late 20th century, the history of animal diseases was not among the prominent topics of historical research. Reasons might include the interdisciplinary character of the topic area and its intersection with veterinary, medical, scientific, economic, environmental, and political contexts. For decades, it was mainly veterinarians or scientific researchers who studied the histories of animal diseases, often with the aim of exploring the historical development of their profession and academic disciplines.[21] In recent years, also in the context of the rise of animal studies, human-animal studies and the 'animal turn', there has been a considerable increase of scholarly interest in the multi-layered relationships between animals and humans. Researchers from various academic disciplines have focused more intensively on the history of animals and animal diseases, including from interdisciplinary perspectives.[22] They have begun to explore the multiple roles of animals in human societies and as historical subjects, as well as the manifold interactions between humans and animals in their political, cultural, social and economic contexts. Researchers have addressed the integration and usage of animals in human activities, their representation in human culture, and changes in human–animal relations in different historical periods. A main focus has been on issues such as the history of livestock husbandry, including in the context of industrialisation, pet keeping, hunting and zoos, especially since the early modern period; other topics include the use of animals in warfare, animal experimentation in research, and animal welfare and rights.[23] Of particular importance to this book is the focus on the exploitation of animals as livestock and its disease-related consequences in the development of a globally networked connected animal industry since the 19th century,[24] as well as the role of animals in scientific research and the historical constructions of the boundaries between animals and humans with regard to diseases.[25] The interest in animal diseases and their history has also been enhanced by the experience of outbreaks of zoonotic diseases since the 1990s, including BSE ("mad cow disease"), SARS and influenza of animals, and the resulting attention of the media and the public.[26] The majority of authors working on the history of animal diseases have so far focused on activities of disease control within the framework of a nation state, a specific colony, a region or a city. Often they put an emphasis on the processes of professionalisation of the veterinary field of work and on the establishment of veterinary institutions. A large part of the research publications deals with questions of animal disease control in the context of Britain, British colonialism, or the US.[27] There are significantly fewer studies on other national, regional or international frameworks, or with a focus on the handling of animal diseases before the 18th century.[28] Joanna Swabe has presented a general investigation into the history of animal disease control in European societies of the 19th and 20th centuries, focusing on "veterinary regimes", which she outlines as "social practices and institutionalised behaviours that have emerged in response to the problem of maintaining animal resources and protecting human health and economy".[29] While her study does not concentrate centrally on international efforts, the focus on animals as a resource to be dependent upon for exploitation emphasised by Swabe[30] is of particular importance also for the present book: for many historical actors, the respective national dependence on a regular trade with livestock was one of the main motivations for addressing the problem area of animal disease control on the international level.

In historical research, scholars have recognised animal diseases increasingly as a problem cutting across different forms of borders and have gradually begun to explore the historical dimensions of this phenomenon. Researchers have studied the history of comparative medicine and its investigations of human–animal relations, including a focus on the handling of zoonotic diseases. In this context, researchers have investigated the history of experimentation on animals, combining perspectives from the history of science and medicine with an analysis of the cultural contexts in Europe.[31] Recently, researchers have more closely addressed the historical dimensions of the links between human and veterinary medicine, analysing the conceptual shift from 'one medicine' to the more comprehensive 'one health' and its focus on biological and ecological aspects of animal and human disease control.[32] Scholars have further investigated the connection between the economic relevance of animal diseases and their impact on public health, in particular the issue of food security. Madeleine Ferrières' study on food-related anxieties and the reactions of researchers and governments has shown how, in various historical contexts from the European Middle Ages to the 19th and early 20th centuries, societies had to weigh their fears of certain animal foods against the omnipresent problem of food shortages.[33] Other researchers have focused on the histories of particular diseases and have analysed control efforts in different national and international contexts. In the case of rinderpest, Clive A. Spinage's detailed work describes outbreaks of the disease and its consequences, as well as responses on a worldwide scale.[34] Amanda Kay McVety's study examines the fight against rinderpest as an international effort with a main focus on the years since the beginning of the Second World War; it is one of the few studies which centrally focuses on the international dimension of the history of animal disease control.[35] Furthermore, scholars have examined historical aspects of specific animal diseases, such as anthrax or bovine tuberculosis, in their scientific, economic, political and societal contexts and in relation to the present.[36] In the case of foot-and-mouth disease, which plays a prominent role in the discussions analysed in the present study because of its high level of contagiousness, its economic relevance and therefore its global significance as a disease cutting across national borders, a comprehensive transregional and transtemporal history of approaches to contain the disease is still missing.[37]

In the last two decades, scholars have put forward new questions concerning the history of animal disease control, among them mainly two directions providing inputs and questions for this study. Firstly, researchers have focused increasingly on veterinary interventions as instruments of colonial rulership in different social and political contexts; most studies concentrate on the British colonies.[38] While the present study does not primarily address veterinary work in colonial contexts, this research provides insights for the analysis of power asymmetries, of strategies of exclusion and of Eurocentric dominance at the interface of the international and imperial levels. Secondly, scholars have begun to focus more closely on the history of animal diseases in context of global economic relations. Mark Harrison has highlighted the links between commercial relations and the spread of contagious diseases in a synthesis from pre-modern times to the present, including the role of animal diseases. He has thereby outlined the complex interplay between governmental regulatory strategies for animal disease control and efforts to maximise commercial gains on the other.[39] Researchers have also focused on the role of animal diseases in the context of commercial relations in specific historical constellations. Of importance for this book are the studies dealing with international discussions on animal diseases as problems of trade in the interwar period.[40] The present study places the history of efforts to address animal disease-related trade problems in the broader context of international discussions and efforts of veterinary organisation which gained momentum since the mid-19th century. Scholars have, in this context, also investigated the problem of potentially sick livestock and infested meat as non-tariff trade barriers in the globalised market of the late 19th and 20th centuries.[41] These studies underline the links between industrialised food production, trade disputes, and public health challenges, providing historical analyses of disputes which were also taken up and problematised in international debates.

The historical actors in veterinary internationalism were in most cases professional veterinarians, physicians or researchers, some of them also politicians, diplomats or civil servants. Almost all were male; women appeared rarely in the role of experts, but more frequently as stenotypists or secretaries. By following the activities of these historical actors, this study is part of recent research on internationalism, international organisations, and the historical actors involved. In the contemporary understanding of the late 19th and early 20th centuries, the term 'international' encompassed a broad spectrum of organisations, associations, professional groups and networks, including intergovernmental institutions.[42] Madeleine Herren has shown how internationalism and international organisations fulfilled a wide range of functions since the middle of the 19th century, and how an analysis of international projects offers insights to better understand the globalisation processes of the time.[43] In a period of intensive national competitiveness, the links between international commitment and national or imperial interests were manifold. As Herren has outlined, international organisations in their multifunctionality included a "power to translate national attitudes into transboundary concepts"[44] because they provided opportunities for cross-border interaction and negotiation.[45] International organisations thus served as hubs for the transfer of information and as forums for the communities of experts who organised their professional exchange across national borders.[46] Governments were interested in international organisations to facilitate trade relations and to benefit from new communication technologies in an increasingly industrialised and globally interconnected world.[47] Both within and beyond the level of institutions, a transnationally connected civil society used international organisations to launch their cross-border activities and expand their social networks. Herren has underlined the multi-levelled impact of these closely connected, overlapping spheres of governmental, professional, commercial and civil society interests and commitment in internationalism.[48] Scholars researching the history of public health and epidemics control in the international contexts of the 19th and 20th centuries have likewise highlighted the interconnections between international, national and imperial concepts and practices of disease control. They have analysed processes of creating anti-disease regulations and establishing institutions in international public health to control and intercept the movements of people across borders. In this context, not only have the activities of international organisations received scholarly attention, as platforms of professional exchange and of multi-layered activities in public health and epidemics control, but also the role and commitment of involved public health experts.[49] Contemporaries used terms such as 'internationalisme médical' or 'internationalisme scientifique' in a broad way to describe a wide range of activities with multiple involved actors, including intergovernmental projects, professional and privately organised transnational initiatives.[50] The present book follows this understanding of internationalism and focuses on the organisations, meetings and initiatives that were considered by contemporaries to be part of the international sphere. Several contemporary reference works on internationalism help us to identify the international associations, congresses, and institutions occupied with the topic area of animal diseases, as well as their members, organisational structures, activities and publications. The first of these reference works were published in the years before the First World War at a time when scholarly research in internationalism was increasing.[51] The League of Nations continued these activities of collecting material on international organisations, conferences and initiatives with its series *Handbook of International Organisations*. The *Handbook* series followed a broad interpretation of the 'international' character of projects and associations: by definition, the series included all international associations, congresses and organisations without commercial interests.[52] The series provided detailed information on member states, officers of the board, objectives, activities, publications, and funding of each international organisation or association, and thus showed the extent of the personal and institutional exchange between the participants of internationalism.[53] A different project of collecting information on internationalism was initiated by the Prussian State Library in Berlin in 1931. The aim was not to provide a reference work for international networking and document the progress of international activities, but rather to establish a comprehensive bibliography of material on international activities in the past and at present. The first (and only) volume appeared in 1939 with a focus on the medical sciences.[54] The editor, librarian and historian Hans Stümke,[55] defined 'international' in a similarly broad way as the League of Nations *Handbook* series did, but stressed that the collection explicitly did not include Jewish associations—not surprising given the place, year and context of publication.[56] This bibliography on medical-scientific internationalism is even more detailed than the League's *Handbook* series. Stümke listed the names of international organisations and congresses in several languages, together with information about the institution and its secretariat as well as publications and conference proceedings. The Berlin bibliography complements the information on veterinary internationalisms given by the League's *Handbooks* with background information on international projects in the field of medicine, public health and veterinary medicine. These contemporary publications provide useful insights for discovering and exploring international scientific-medical activities until the late 1930s. For researchers, the situation of historical sources on the activities of international cooperation in the field of animal diseases, especially in the period before 1945, still involves difficulties on several levels. Material about transnational professional networks or non-official international initiatives is often difficult to locate. Intergovernmental organisations such as the League of Nations usually maintained own institutional archives; frequently they also produced printed documentation of their activities and edited series of publications. In case of the OIE, the archival material for the period before 1945 is more fragmented. The source material on veterinary internationalism therefore needs to be complemented by the records of several national archives and of cooperating international organisations. For the present study, especially the material preserved in the French Archives Diplomatiques documenting administrative processes of international offices located in France has provided important insights. For the period of the Second World War in particular, the material preserved in the Politisches Archiv des Auswärtigen Amts (Political Archive of the Federal Foreign Office) in Berlin and in the Schweizerisches Bundesarchiv (Swiss Federal Archives) in Berne, because of the involvement of Swiss actors in international organisations, have been helpful. Furthermore, the documents published by the international organisations and associations themselves, in particular the reports, minutes of meetings and conference proceedings, provide insights into the development of their international projects, the institutional transformations and the networking activities as well as on the concepts and practices discussed by their members.

Approaching the historical globality of animal diseases

In November 1937, the epidemiologist Frank G. Boudreau published an essay in the *American Journal of Public Health* on the global spread of contagious diseases and the international cooperation in the field of public health and hygiene. Boudreau had before worked for more than a decade in senior positions in the Health Section of the League of Nations Secretariat and was known as an advocate of international cooperation to address common problems of public health.[57] His essay reflected the contemporary perception of a time of unprecedented technical, medical and scientific progress. In Boudreau's view, the new global situation of radically reduced spatial and temporal distances had profound consequences for the spread of contagious diseases and for international responses to the problem:

> In comparison with the march of civilization, the rapidity of its advance [of international cooperation] is seen to leave something to be desired. For we are living in a world which is shrinking visibly and steadily, year by year and day by day.... This world physical shrinkage has brought in its train many by-products. Isolation is no longer possible anywhere. Eternal vigilance must be exercised to prevent the spread of disease.... But the complexity of the world is due to the increasingly dense network of social, economic, and political relationships which caused the murder of an archduke in far away Sarajevo to bring on a war which affected every single inhabitant of the globe.... But the hard facts of the machine age and the industrial revolution cannot be denied. Coöperate [sic] or perish are the two horns of the dilemma. There is no alternative, no other path of escape.—I believe that the world today is turning, perhaps very slowly, toward the alternative of cooperation. The progress of international cooperation in health matters illustrates the trend in that direction.[58]

Boudreau's reflection about the consequences of a speed- and progress-oriented interconnected world provides some insights for the analysis of international projects to contain global animal disease challenges. In his description, the perception of global spatial and temporal connectedness of what he called the "machine age"—especially the "world physical shrinkage"—went hand in hand with the creation of new social and political connections on many different levels and in an accelerated pace. This combined process resulted in what he perceived as a growing "complexity of the world". One of its downsides was a more spatially extensive and quicker spread of contagious diseases across borders and around the world. He argued that intensive and continuous efforts of international cooperation against communicable diseases were a necessity for human survival, and they were without any real alternative—"coöperate or perish". These international efforts were, in his view, important to start new forms of "eternal vigilance" to improve disease prevention on a global level. In Boudreau's view, such efforts of international cooperation to control the spread of infectious diseases related primarily, but not exclusively, to activities at the intergovernmental level. Reflecting the author's goals and professional background, the essay showed his confidence in efforts of international cooperation as the de facto only way to limit the ubiquitous problematic consequences of progress-oriented global interconnections.

In studies of the history of epidemics and epizootics, research on border constructions and policies as disease control measures has highlighted the importance of spatial organisation in relation to the containment of contagion.[59] Alison Bashford has described the history of world health policies as "merged formations of colonial, national, and 'world' politics, played out on specific local ground" with overlapping spaces of regulation.[60] She has highlighted the importance of both territorial and deterritorial regulatory measures, with borders in context of public health measures being "more than abstract lines on maps, but a set of practices on the ground".[61] Patrick Zylberman has underlined the link between the political and the epidemiological objectives of national governments as well as the role of disease control in discourses of 'civilisation' and the geopolitical interests of nation states.[62] Scholars in the field of border studies have provided further perspectives for investigating the function and the construction of borders in epidemiological contexts. They have highlighted the conceptual shift in the understanding of borders from territorial lines towards practices and processes.[63] In this context, researchers have analysed the diversity and multiplicity of borders, their shifting character, and their functions of channelling and structuring mobilities. A closer focus on the construction of border policies thus can give insights into power asymmetries and inequalities of a globally oriented capitalist system.[64] Historians have also addressed the processes of defining spaces as specifically demarcated areas of state sovereignty that were subject to increased policing and administration especially since the late 18th century. These activities included the introduction of border-related measures against the spread of epidemics and epizootics. Borders were constructed, defined and revised by historical actors, and borderland regions were characterised by multiple entwined social, economic, political and cultural interrelations.[65] As this book argues, in the history of epizootics containment—the term itself referring to the idea of spatial limitation as a measure—borders manifested in several ways. They were political-territorial and material lines of demarcation, and instruments of disease control combined with sanitary veterinary control practices that were defined, constructed, stabilised, destabilised, deconstructed and redefined in specific historical constellations. Sanitary boundaries did not necessarily have to coincide with political boundaries, but were also established regionally and provisionally for a limited period of time. The practices of regulation used at sanitary border stations were developed to intercept the spread of animal diseases, spatially, temporally and materially, in order to allow the cross-border transfer of animals for travel and trade. Borders also appeared as biological boundaries, for example between animal species, and as conceptually constructed demarcation lines. The understanding that boundaries between species could be crossed in case of certain diseases was central for the historial scientific work on zoonoses at the interface of veterinary medicine and public health. The establishment of borderlines thus included the processual dimension of creating demarcations, of establishing scientific categories, and of substantiating discourses of exclusion and inclusion. Spatial-territorial demarcations were also needed as frameworks to assess, organise and monitor the cross-border travel of potential disease carriers. As Patrick Zylberman has outlined, "the border took on the role of strategic keystone and watchtower for health".[66] The most important material, temporal and spatial border-related practices in use for animal disease control in the 19th century, heavily discussed on international platforms, were the veterinary sanitary measures of inspection, quarantine, isolation of suspected animals, disinfection of animals and of the means of transport, and the introduction of official health certificates for the documentation of cross-border passages.

However, for an effective control of epidemics and epizootics in a globalised world, the construction and expansion of sanitary borderlines equipped with control practices was not sufficient. Contemporary experts were very aware of this problem. Epizootics had a tendency to spread across entire continents—a characteristic that had not changed significantly for centuries, also not in modern times.[67] Even the most technically and scientifically sophisticated border regimes were not always able to stop the spread of highly contagious animal diseases such as foot-and-mouth disease or rinderpest, especially given the increasing dependencies on livestock imports in many countries. The processes of disease spread were complex, involved many potential carriers and varieties of pathogens, and they depended on the changing human activities in their political, economic and social contexts. Wild animals as disease carriers, especially birds or rodents, were difficult or impossible to be controlled in their mobility. Furthermore, especially in times of war, territorial sovereignty, which was the framework for border-based epizootics control, could be called into question and the implementation of territorial-based control practices became more difficult to be maintained. Borderlines, including sanitary ones, became more mobile or obsolete as a result of war-related territorial shifts and of changes in the organisation of border areas. At international meetings, therefore, historical actors discussed not only the question of how to cooperate in organising sanitary veterinary borderlines and coordinate the crossing of borders for animals and goods. They increasingly debated how to develop new tools for disease control that took into account the problem of the deficiencies of borders and their associated practices, and underlined the need for further approaches to handle acute outbreaks and the spread of animal diseases. Inspired by the international discussions on human epidemics control and the idea of expanding epidemiological intelligence, the concept of a centralised system of information transfer with regard to epizootics increasingly came into focus.[68] International meetings debated intensively about what exactly animal disease information would be, what it consisted of, and how it would have to be generated, organised and distributed in order to be effective for disease control across borders. The territorial organisation of animal disease control, in conjunction with veterinary measures at borders, did not lose their importance. But the ideas of a regular international exchange and transfer of information on contagious animal diseases and of the introduction of common international veterinary standards of disease control gained momentum. In addition, since the late 19th century, experts gradually developed ideas for transboundary vaccination programmes. It was the permeability of borderlines that called for new concerted international activities to be implemented within and beyond borders.

As one can see in Boudreau's essay, similar to today, the spread of epidemics and epizootics was in the interwar period regarded as an indicator of the problematic sides associated with the progress-oriented processes of what today is called globalisation. Scholars have tried to make the often analytically vague concept of globalisation useful for historical analyses, emphasising the plural character of 'globalisations' as historical processes, and the transformative effects of these processes with regard to the social interaction of people.[69] Further, they have underlined that historicising globalisation includes searching for an awareness of processes of global densification among historical actors.[70] The economic dimension of globalisation processes, with the increasing interconnectedness of markets and flows of goods, including animals and animal products, played an important part with regard to the spread of animal diseases. In studying the links between the spread of epidemics and epizootics and globalisation processes, researchers have observed that the latter did not necessarily result in a perforation or dissolution of borders on a global scale. As Alison Bashford put it, "the 'age of globalization'—the world post-HIV/AIDS, post multi-drug resistant TB, West Nile Virus and SARS—might be global in terms of disease spread, but is also characterised by increasingly intense regulation at national borders".[71] Discussing the connection between processes of globalisation and the spread of epidemics, Mark Harrison has pointed out to consider more closely how the social, economic and ecological changes produced by humans profoundly influenced the emergences and spread of diseases on a global level, and how diseases themselves acted as agents of global change.[72] Indeed, the transformations of the 19th and 20th centuries and their ramifications on a global and a local level undoubtedly shaped the spatial and temporal patterns by which animal diseases spread throughout the world. It is, however, important to consider that the emergence and spread of animal diseases were not only connected to the globalisation of trade through the technical progress since the 19th century, but animal diseases were present in all historical constellations and dependent on the changing economic, ecological, political and social conditions.

This study follows methodological reflections in context of global history to explore historical connections and interactions in international efforts to contain animal diseases. Sebastian Conrad has pointed to the tendency of researchers in global history to foreground historical connections and mobilities; Jürgen Osterhammel has similarly noted a "mobility bias" in global history research.[73] If one of the concerns of global history approaches is to dismiss teleological narratives of modernisation,[74] focusing on the topic of animal disease control in its entangled cross-border contexts can help to reflect on this aim. The topic invites to problematise the idea of historical connectedness—to think on how and by whom historical connections in the global context of animal disease containment were created and structurally consolidated, and how they got reshaped, fragmented or vanished. It can throw light on the manifestations of power and hierarchies expressed in the processes of initiating and substantiating connections or disconnections and of dealing with their consequences. Thus it can, as this book suggests, help to reveal disparities and inequalities which societies and individuals faced when confronted with potentially harmful globally connected problem areas such as contagious animal diseases. The understanding of the entangled globality of animal diseases in this book, therefore, includes the potentially almost planetary dimension of the spread of pathogens, while emphasising the multitude of global-local uncertainties and implications associated with animal diseases and the efforts of their control. In such a historical approach to animal disease control, the aspect of mobility is a central element, as is a focus on the contemporary perceptions of connections with the idea of condensed distances in spatial and temporal terms due to the globalisation of trade and travel. International efforts to address animal disease problems, as this study shows, were thus frequently aimed at reducing or eliminating the uncertainties associated with the mobility of pathogens. An animal disease such as anthrax travelled globally, but it could also, for instance, cause local outbreaks over a long period of time through pathogens remaining in the soil. Even if an animal disease did not break out eventually at a certain place, the fear of its spread could lead to interventions and reactions from authorities, disease experts, the media and the public. Global connections, this book shows, were in the case of animal diseases always highly ambiguous and subject to frequent change: the problems of animal diseases manifested in a sort of grey zone of simultaneous temporal and spatial presence and absence, both locally and globally. The forms and patterns of the mobility of diseases were often not clear since pathogens adapted to their host and vector animals, as well as to human behaviour. The links of animal diseases to human activities therefore created further factors of uncertainty. Pathways of transmission could be, for instance, direct contact, or droplet infection, inhalation of dust, the contagion via vector animals, contaminated food or the contact with animal products. The spread, stagnation, disappearance and re-emergence of animal diseases were only partially or often not at all predictable; they sometimes changed their appearance and aggressiveness, adapted or remained hidden for years, or emerged in different animal species. Treatments and preventive measures effective beyond borders, especially in the form of vaccinations, were in the past—and are also today—often expensive and difficult to be applied effectively on animals on a large scale, or they were simply non-existent. While the mobility of diseases thus played an essential role in their perception as a threat and with regard to the development of anti-disease measures, less mobile people or animals could be directly or indirectly affected by the spread of pathogens. Similar to other problems endangering health and prosperity, contagious animal diseases, especially zoonotic ones, potentially included an immediacy and directness that could become existential for human beings. The entangled globality of animal diseases thus was—and still is—linked to societies in total, and at the same time directly to the lives of individuals, including a local, even micro-level with an individual dimension. Conceptually, the problem transcends the dichotomy between connection and disconnection with its overlapping layers of potential local-global threats. The simultaneity of the processes of global connections, reconnections and disconnections is thus a component of the uncertainties associated with animal diseases. The manifestations of global disease connections appeared as real and potential, absent and present, planetary and local. The focus of this book is therefore not on tracking the worldwide spread of historical epizootics events with their mobility of pathogens. Rather, it deals with the historical perceptions of disease entanglements and the development of responses, the interpretations of globality associated with them, and the consequences of such conceptualisations for addressing global problem areas.

"An international pool of ideas": expert internationalism and animal diseases

In 1931, the British political scientist H. R. G. Greaves, who during the 1920s had studied in Geneva and since 1930 worked at the London School of Economics, published a study on the new international role of "experts and technicians", as he called them, in particular in the committees of the League of Nations.[75] He outlined the increasing significance of expertise in context of international relations:

> The growing need for an international reply to questions soluble only by international action, and secondly, the increasing authority of the expert administrator at home, both combine to stress the importance of these international committees of technicians.... Members are not government representatives meeting to drive a bargain and gain the

> utmost advantage which the power of their country can extort.... By this means, an informal co-operation is possible. It results in a continual exchange of ideas. It sends views back and forth from one country to another, creating, as it were, an international pool of ideas.[76]

Greaves described the creation of what he called an "international pool of ideas" built up by experts as an important and innovative element of international cooperation in his time, and highlighted the professional authority of experts which added to their legitimacy to be active on the international level. The professional connections of the experts and their ability to pursue informal networking and cooperation activities contributed, in his opinion, to their increasingly important role on the international stage. This idea of the figure of the expert included a delineation from political decision makers and from the public. Greaves thus underlined an idealised image of the role of experts in processes of policymaking, in this case on international platforms, that was popular at the time—experts cooperating with each other on a professional level to establish technical, fact-based and scientifically oriented solutions with the aim to address complex and internationally relevant problems, while supposedly remaining politically neutral and outside of international conflicts. The creation of this dynamic "international pool of ideas", as Greaves called it, was indeed of central importance for the work of experts aiming to address global problem areas such as the spread of animal diseases. Scholars have in recent years focused increasingly on how experts and groups of experts have created, shaped and transformed the platforms and formats of internationalism.[77] As already Greaves described, the integration of experts in governmental tasks, and in a further step into matters of foreign policy and in international organisations, gained momentum with the expanding and differentiating of bureaucracies on the level of nation states. Governments were in need of and relying on experts' professional advice concerning an increasingly complex range of problem areas. On the international level, they needed experts as professional advisors and representatives in conferences and in international organisations. Additionally, many professional groups began to form networks and create associations and platforms of exchange beyond national borders. Experts became thus increasingly organised in international associations, and at the same time involved in international governmental relations with regard to areas such as cross-border epidemics and animal disease control. After the First World War, the League of Nations and other international institutions and organisations structurally integrated experts in key positions to create new networks, formats and opportunities of international cooperation. In her study on the economic work of the League of Nations, Patricia Clavin has stressed the—sometimes confusing—shifting agendas, manifold areas of activity, multiplicity of actors in the commissions and subcommissions, and the interrelated personal networks as part of the League's innovative foodprint. Analysing the case of the League's Economic and Financial Organisation, she has shown how a network of multiple actors, among them many technical experts, 'reinvented' the League by searching for innovative responses to urgent problems caused by the worldwide political and economic crisis.[78]

The category of experts in this study refers to historical actors with an academic or practical background in the field of veterinary medicine or related fields of work. The concept of epistemic communities,[79] developed to examine the role of expertise in policy-relevant areas, provides some analytical elements which are helpful to focus on the question of how expert groups interacted in the case of veterinary internationalism. The concept indicates the connecting elements of epistemic groups, their "shared set of normative and principled beliefs",[80] and the policy-shaping forms of communication of expert knowledge.[81] In this study of veterinary internationalism, the aim is not to trace the creation of an epistemic community of veterinary scientists, nor to provide a story of modern expert rule. Rather, the book focuses on how experts as historical actors explored and used their scope of action at the international level when confronted with problems of a global dimension. As mentioned earlier, similar to other arenas of internationalism, veterinary experts frequently held leading positions in national and governmental administrations, in academia or in research institutions parallel to their international activities; they attended international meetings thus in overlapping roles.[82] The image of the expert figure as a politically neutral outsider committed to providing specialised technical knowledge for decision makers is therefore insufficient for understanding their role at the international level. As this book shows, experts active in veterinary internationalism indeed gradually built up, as Greaves called it, an "international pool of ideas"[83] on the subject of animal diseases which was central for the development of any form of cross-border approach to the problem area. They organised and provided suggestions for international cooperation and for joint activities, and communicated their projects to national decision makers and to their professional networks. The general public usually played a minor role in this context. The work of expert groups involved in veterinary internationalism was shaped by common convictions, primarily the professional discourses provided by the dominant European veterinary scientific communities. This also resulted in problems because experts, revolving around their own professional environment, could fail to recognise gaps and shortcomings. Most experts on the subject of animal diseases present at international meetings did not question the universal applicability of European scientific concepts for their work. The factor of their national identity was important, and national loyalties or rivalries often contributed to tensions within the expert community. Many were professionally involved in the decision-making processes on animal disease control on a national level, and at international forums they highlighted the relevance of their respective national policy approach. Platforms of veterinary internationalism were highly competitive places, and the discussions included the tendency to make comparisons of the respective national efforts in disease control and scientific excellence.

It is difficult to measure the effects of the activities of experts at international platforms in the late 19th and first half of the 20th centuries, especially if their meetings did not produce clearly visible results such as an international arrangement.[84] As mentioned before, this study uses the idea of 'organising trust' to approach the multi-levelled activities of experts on platforms of veterinary internationalism. In international relations research, scholars following a rational perspective have focused on the idea of trust as a strategic resource and a calculation of decisions to underscore national interests.[85] Other scholars have emphasised the social, emotional and psychological dimensions of trust and its complex and multi-layered role in the development of international relations.[86] With regard to historical research, Ute Frevert has explored "questions of trust"[87] from various perspectives. Frevert has outlined how trust was and is an "obsession" of modern societies, and how trust has become an important part of the relationship between the state and its citizens, as people needed to develop trust in the functioning of institutions and their regulatory powers.[88] Geoffrey A. Hosking has described trust as important "ingredient in this web of interdependence"[89] of social life. "Trust networks", in his analysis, operated with a certain stability due to the commitment and mutual ties of their members, although not without conflict; furthermore, the participants showed a tendency to exclude those who were defined as others.[90] In context of a history of veterinary internationalism, the analytical focus on 'organising trust' takes into view how historical actors approached the topic area of animal diseases, trying to overcome the obstacles and tensions which they were confronted with in context of their work. The aspect of 'organising' reflects the processual element in the development of concepts, practices and institutions of veterinary internationalism. Many of these activities originated from the regulatory repertoire developed in subject areas of internationalism since the 19th century. This study identifies as core part of 'organising trust' in veterinary internationalism two main elements. Firstly, from the beginning, the strategic goal of many participating experts was to commit governments to binding collective action in the form of international conventions and the foundation of new institutions. The idea was not only to incite a more active state interventionism against animal diseases in general, but also to advance the concept of uniform national regulatory frameworks to facilitate coordinated approaches of animal disease control. Internationally uniform regulations and standards would, according to the assumption of experts, make the respective disease situation in the trading countries more ascertainable and more trustworthy, and therefore the trade with animals more secure. Experts intended, additionally, to initiate joint international projects for the containment and prevention of animal diseases. Secondly, experts advanced the idea of establishing a transparent and permanent international system of information transfer on animal diseases. As this study shows, international meetings discussed and organised techniques to create trustworthiness by organising a continuous and reliable flow of high-quality information from national and regional sanitary administrations. This goal of increasing transparency and trustworthiness through a regular and systematic exchange of information concerning ideally the worldwide animal disease situation again included the idea of creating international standards and norms to allow a scientifically based assessment of the animal disease situation. The idea of building up relations and structures of cooperation and communication based on trust was conceptually intertwined with control measures such as the practices of veterinary control at borders. Control measures had their limits, which was why relations based on trust were considered by many as necessary to advance steps against epizootics. The organisation of trustful relations discussed at the platforms of veterinary internationalism was shaped by power asymmetries, scientific rivalry and national competition on several levels between different governments and experts. The common scientific principles of expert communities and the master narrative of European scientific and technical progress were elements that supported the processes of 'organising trust'. Still, as the historical actors noticed, it was often in any case challenging to initiate and realise projects of collaboration addressing animal disease problems. The approaches offered by international law, in particular the negotiation of multilateral conventions, frequently proved to be difficult in terms of negotiation and implementation. The processes summarised by 'organising trust' thus also served to compensate for such limitations through the activities of experts and their communities. As the discussions on international platforms showed, many of the participating historical actors considered the continued development of trust-based international cooperation and coordination not as an option, but as an urgent and indispensable necessity.

Chapter outline

The following chapters of this book trace the projects of veterinary experts on international platforms and institutions in the period from the 1860s to the early years after the Second World War chronologically, taking into account parallel and overlapping developments. After this introductory part, Chapter 2 provides a historical contextualisation of the multiple economic, political and social problems associated with animal diseases and their spread in the mid- and late 19th and early 20th centuries. During this period, groundbreaking developments in agriculture, industrialisation, international trade and scientific research created new and more globally interlinked problem areas related to animal diseases, as well as new opportunities of disease control. In Europe, governments addressed the area of animal disease containment by integrating the issue more intensively into the tasks of public authorities and by creating institutions for the professional training of veterinary experts. In the late 19th century, the question of adequate measures tackling the global spread of animal diseases became increasingly discussed, in particular since animal diseases posed a problem of international trade in animals and animal products in connection with public health concerns. The chapter follows these intertwined problem areas and examines how the globalised spread of animal diseases created—in the eyes of veterinary experts—a need for new forms of permanent and more sustainable international cooperation.

Chapter 3 examines discussions at the early platforms of veterinary internationalism from the 1860s to the First World War. The main focus is on the series of international veterinary congresses, which offered veterinary experts a new platform to develop strategies addressing animal disease problems in their cross-border dimension. The first international veterinary congress in Hamburg took place in 1863 as a meeting of experts; it was the starting point for a growing international debate on the subject of animal disease control. In 1872, an international conference in Vienna attempted to advance joint international regulations against rinderpest, which threatened the livestock sector in Central Europe. Further international veterinary congresses followed until 1914, when the outbreak of war abruptly interrupted international cooperation efforts. As a result of the debates at international platforms, this chapter argues, experts developed an instrumentarium of international concepts and practices of animal disease control that also addressed the problem of their cross-border spread. But the implementation in practice remained difficult, mainly because of the hesitant attitude of many governments to participate in binding international agreements.

Chapter 4 deals with the establishment of a new international veterinary organisation after the First World War. The chapter shows how a new era of international cooperation in the fight against animal diseases began in the early post-war years, reflected in new institutional structures and the integration of experts within international organisations. The chapter focuses in particular on the Conférence Internationale pour l'Etude des Epizooties taking place in Paris in 1921. The participants of the conference agreed on the creation of a permanent international office to deal with the topic area of epizootics containment. The new international office, the OIE, started its work in 1927. The chapter examines the directions in post-war international veterinary work and addresses the expectations of participants to approach animal disease problems through a new international institution. The conference of 1921 and the founding history of the OIE, the chapter argues, revealed the tensions between a European-dominated international focus on the problem area and the global challenges created by animal diseases in an interconnected world.

Chapter 5 examines the initiatives of several cooperating international organsations addressing specific issues of animal disease control in the period before the Second World War. The projects also revealed the complexity of finding responses towards animal disease problems at the interface of human and veterinary medicine. A joint initiative of the League of Nations and the ILO dealt with anthrax as an occupational disease affecting factory workers, and a second, smaller project of the League in collaboration with the Pasteur Institute network focused on investigating the worldwide practices of anti-rabies vaccination. Furthermore, the League aimed to facilitate trade relations through the establishment of international conventions on animal disease control. The chapter analyses the work of historical actors in the different projects and shows that the many obstacles faced by veterinary experts prompted them to think about flexible practices and concepts—experimenting with different layers of international work became a key to advancing cross-border anti-disease activities.

Chapter 6 deals with the work of the Office International des Epizooties (OIE) in Paris from the late 1920s until the end of the Second World War and in the early post-war years. The staff of the OIE and its associated veterinary experts began to build up a new permanent international system of information transfer for the control and prevention of animal diseases. The chapter explains how, in the face of increasing international tensions during the 1930s, the OIE tried to advance international efforts to address the spread of animal diseases. The OIE continued its work in occupied Paris during the Second World War, despite manifold problems posed by the German occupying authorities which aimed to secure its war-related livestock supply chains. The chapter argues that, especially in the immediate post-war period, the need for both a more global approach and a conceptual change with regard to international animal disease control became increasingly apparent: the OIE had to discuss new concepts and orientations that needed to go beyond organising the transfer of information and to encompass a more global outlook on animal disease control.

Notes

1 About the rinderpest outbreak in Belgium in 1920: Joseph Ricard, *Les Origines de la Première Conférence Internationale des Epizooties, Fièvre Aphteuse, Peste Bovine* (Paris: de Chaix, 1921), 7; Marc Mammerickx, 'La Peste Bovine, Jules Bordet et le Centre Sérumigène de Cureghem', *Annales de Médecine Vétérinaire* 147 (2003): 200–3; Amanda Kay McVety, *The Rinderpest Campaigns: A Virus, Its Vaccines, and Global Development in the Twentieth Century*, (Cambridge: Cambridge University Press, 2018), 13.

2 République Française, Ministère de l'Agriculture, ed., *Conférence Internationale pour l'Etude des Epizooties. Paris (25–28 Mai 1921). Compte Rendu Sténographique* (Paris: Imprimerie Nationale, 1921), 21–2; Clive A. Spinage, *Cattle Plague: A History* (Dordrecht and London: Kluwer, 2003), 211–12; McVety, *The Rinderpest Campaigns*, 13–14.

3 Ricard, *Origines de la Première Conférence Internationale des Epizooties*, 4–5.

4 For the conference minutes: République Française, Ministère de l'Agriculture, *Conférence Internationale pour l'Etude des Epizooties 1921*. About the conference and the conference preparations see also Chapter 4, as well as: Mark Harrison, *Contagion: How Commerce Has Spread Disease* (New Haven: Yale University Press, 2012), 226–7; McVety, *The Rinderpest Campaigns*, 13–5, 33.

5 World Organisation for Animal Health (OIE), 'International Agreement for the Creation of an Office International des Epizooties in Paris, done in Paris, January 25, 1924', accessed May 15, 2021, https://www.oie.int/about-us/key-texts/basic-texts/international-agreement-for-the-creation-of-an-office-international-des-epizooties/.

6 World Organisation for Animal Health (OIE) and Food and Agriculture Organization of the United Nations (FAO), 'Final Report of the Joint Committee on Global Rinderpest Eradication, May 2011', accessed May 15, 2021, http://www.oie.int/doc/ged/D10943.PDF.

7 On the global fight against rinderpest leading to the eradication of the disease: McVety, *The Rinderpest Campaigns*.

8 World Organisation for Animal Health (OIE), 'One Health', accessed May 28, 2021, http://www.oie.int/en/for-the-media/onehealth/. On COVID-19 in animals: World Organisation for Animal Health (OIE), 'COVID-19', accessed May 30, 2021, https://www.oie.int/en/what-we-offer/emergency-and-resilience/covid-19/.

9 About the spread of animal diseases in the global contexts of the 19th century see also: Harrison, *Contagion*, 211–13.

10 For the connection between industrialised animal farming and the development of diseases: Harrison, *Contagion*, 279.

11 Martin H. Geyer and Johannes Paulmann, eds., *The Mechanics of Internationalism: Culture, Society, and Politics from the 1840s to the First World War* (Oxford, New York: Oxford University Press, 2001). For research on the history of internationalisms: Glenda Sluga and Patricia Clavin, eds., *Internationalisms. A Twentieth-Century History* (Cambridge: Cambridge University Press, 2017).

12 Alison Bashford, '"The Age of Universal Contagion": History, Disease, and Globalization', in *Medicine at the Border. Disease, Globalization and Security, 1850 to the Present* (Houndmills, Basingstoke: Palgrave Macmillan, 2006), 6.

13 Dipesh Chakrabarty, *Provincializing Europe: Postcolonial Thought and Historical Difference* (Princeton, NJ: Princeton University Press, 2000).

14 Bashford, '"The Age of Universal Contagion": History, Disease, and Globalization', 6.

15 About the notion of 'civilisation' in context of the history of international public health: Patrick Zylberman, 'Civilizing the State. Borders, Weak States and International Health in Modern Europe', in *Medicine at the Border. Disease, Globalization and Security, 1850 to the Present*, ed. Alison Bashford (Houndmills, Basingstoke: Palgrave Macmillan, 2006), 21–40; Patrick Zylberman, 'Making Food Safety an Issue: Internationalized Food Politics and French Public Health from the 1870s to the Present', *Medical History* 48, no. 1 (2004): 5.

16 The PAHO started its veterinary public health programme in 1949 in cooperation with the WHO, see: Pan American Health Organization, 'Veterinary Public Health – PAHO/WHO', accessed April 1, 2021, http://www.paho.org/en/topics/veterinary-public-health. About the history of the PAHO: Marcos Cueto, *The Value of Health: A History of the Pan American Health Organization* (Washington, DC: Pan American Health Organization, 2007).

17 Madeleine Herren, *Hintertüren zur Macht: Internationalismus und modernisierungsorientierte Aussenpolitik in Belgien, der Schweiz und den USA, 1865–1914* (München: Oldenbourg, 2000).

18 See also: Harrison, *Contagion*, 212.

19 The Polish epidemiologist Ludwik Rajchman (1881–1965) was one of the leading figures of 20th century medical internationalism, working as director of the League of Nations Health Organisation until 1938. After 1945, he became representative of the UNRRA in Poland and one of the founders of UNICEF. For a biography: Marta Aleksandra Balinska, *For the Good of Humanity: Ludwik Rajchman, Medical Statesman* (New York: Central European University Press, 1998).

20 Rajchman, in: Société des Nations, Comité Economique, Sous-Comité d'Experts en Matière de Mesures de Police Vétérinaire. Première Session Provisoire, Troisième Séance, January 31, 1928, 8. LON R2774, 10D/2049/272 (translation: CK; original: "La confiance est à la base de ces travaux; il faut l'organiser, car il y a un principe permanent, valuable également en hygiène humaine, qui est que l'on doit prendre uniquement des mesures d'une efficacité éprouvée et en proportionner la rigueur aux menaces de contagion").

21 See: Joanna Swabe, *Animals, Disease, and Human Society. Human-Animal Relations and the Rise of Veterinary Medicine* (London and New York: Routledge, 1999), 9. An important work is: Calvin W. Schwabe, *Veterinary*

Medicine and Human Health, 2nd ed. (Baltimore: Williams & Wilkins, 1984). The first director of the OIE, Emmanuel Leclainche, published a broad overview of veterinary history: Emmanuel Leclainche, *Histoire de la Médecine Vétérinaire* (Toulouse: Office du Livre, 1936). Another former director of the OIE, Jean Blancou, included in his exploration of the history of animal diseases the contexts of veterinary public health: Jean Blancou, *Histoire de la Surveillance et du Contrôle des Maladies Animales Transmissibles* (Paris: OIE, 2000).

22 For instance: Kristin Asdal, Steve Hinchliffe, and Tone Druglitrø, eds., *Humans, Animals and Biopolitics: The More-than-Human Condition* (Abingdon, New York: Routledge, 2017). For the 'animal turn': Harriet Ritvo, 'On the Animal Turn', *Daedalus* 136, no. 4 (2007): 118–22.

23 For an introduction into 'animal histories': Dorothee Brantz, ed., *Beastly Natures: Animals, Humans, and the Study of History* (Charlottesville: University of Virginia Press, 2010). For an overview with perspectives from different academic disciplines: Roland Borgards, ed., *Tiere. Kulturwissenschaftliches Handbuch* (Stuttgart: Metzler, 2016); Philip Howell and Hilda Kean, eds., *The Routledge Companion to Animal-Human History* (London, New York: Routledge, 2018).

24 For an overview: Lukas Nieradzik, 'Geschichte der Nutztiere', in *Tiere. Kulturwissenschaftliches Handbuch*, ed. Roland Borgards (Stuttgart: Metzler, 2016), 121–9.

25 Abigail Woods et al., *Animals and the Shaping of Modern Medicine: One Health and Its Histories* (London, New York: Palgrave Macmillan, 2018).

26 On influenza in interdisciplinary research, relating history and the present, see: Ann Herring and Alan C. Swedlund, eds., *Plagues and Epidemics: Infected Spaces Past and Present* (Oxford, New York: Berg, 2010). In detail on the links of contagious and zoonotic diseases in relation to globalisation: Harrison, *Contagion*, 245–75.

27 For example: Alan L. Olmstead and Paul W. Rhode, *Arresting Contagion: Science, Policy, and Conflicts over Animal Disease Control* (Cambridge, Mass.: Harvard University Press, 2015); Abigail Woods, 'Is Prevention Better than Cure? The Rise and Fall of Veterinary Preventive Medicine, c. 1950–1980', *Social History of Medicine* 26, no. 1 (2013): 113–31; Abigail Woods, 'From Practical Men to Scientific Experts: British Veterinary Surgeons and the Development of Government Scientific Expertise, c. 1878–1919', *History of Science* 51, no. 4 (2013): 457–80; Alan L. Olmstead, 'The First Line of Defense: Inventing the Infrastructure to Combat Animal Diseases', *The Journal of Economic History* 69, no. 2 (2009): 327–57; Neil Pemberton and Michael Worboys, *Mad Dogs and Englishmen: Rabies in Britain, 1830–2000* (Houndmills, Basingstoke: Palgrave Macmillan, 2007); Keir Waddington, 'To Stamp out "so Terrible a Malady": Bovine Tuberculosis and Tuberculin Testing in Britain 1890–1939', *Medical History* 48, no. 1 (2004): 29–48; Abigail Woods, *A Manufactured Plague: The History of Foot-and-Mouth Disease in Britain* (London: Earthscan, 2004); Anne Hardy, 'Professional Advantage and Public Health: British Veterinarians and State Veterinary Services, 1865–1939', *Twentieth Century British History* 14, no. 1 (2003): 1–23.

28 On the early modern period see for instance: Katharina Engelken, Dominik Hünniger, and Steffi Windelen, eds., *Beten, Impfen, Sammeln. Zur Viehseuchen- und Schädlingsbekämpfung in der Frühen Neuzeit* (Göttingen: Universitätsverlag Göttingen, 2007); Madeleine Ferrières, *Sacred Cow, Mad Cow: A History of Food Fears* (New York: Columbia University Press, 2006); Blancou, *Histoire de la Surveillance*. On Germany: Dominik Hünniger, *Die Viehseuche von 1744–52. Deutungen und Herrschaftspraxis in Krisenzeiten* (Neumünster: Wachholtz Verlag, 2011). On France: François Vallat, *Les Bœufs Malades de la Peste. La Peste Bovine en France et en Europe, XVIIIe-XIXe Siècle* (Rennes: Presses Universitaires de Rennes, 2009).

29 Swabe, *Animals, Disease, and Human Society*, 8.

30 Swabe, *Animals, Disease, and Human Society*, 5–9.
31 Woods et al., *Animals and the Shaping of Modern Medicine*; Anne Hardy, 'Animals, Disease, and Man: Making Connections', *Perspectives in Biology and Medicine* 46, no. 2 (2003): 200–15; Lise Wilkinson, *Animals and Disease. An Introduction to the History of Comparative Medicine* (Cambridge and New York: Cambridge University Press, 1992). On the history of experiments with animals: Anita Guerrini, *Experimenting with Humans and Animals. From Galen to Animal Rights* (Baltimore: Johns Hopkins University Press, 2003).
32 Research project 'One Medicine? Investigating Human and Animal Disease', accessed March 5, 2021, http://www.kcl.ac.uk/artshums/depts/history/research/proj/onemedicine.aspx. See also: Michael Bresalier, Angela Cassidy, and Abigail Woods, 'One Health in History', in *One Health. The Theory and Practice of Integrated Health Approaches*, ed. Jakob Zinsstag et al. (Wallingford, Boston: Cabi, 2015), 1–15; Woods et al., *Animals and the Shaping of Modern Medicine*.
33 Ferrières, *Sacred Cow, Mad Cow*, 4. For questions of food security in diverse historical contexts see also: David F. Smith and Jim Phillips, eds., *Food, Science, Policy, and Regulation in the Twentieth Century: International and Comparative Perspectives* (London and New York: Routledge, 2000).
34 Spinage, *Cattle Plague*.
35 McVety, *The Rinderpest Campaigns*.
36 Susan D. Jones, *Death in a Small Package. A Short History of Anthrax* (Baltimore: Johns Hopkins University Press, 2010); Philipp Sarasin, *Anthrax: Bioterror as Fact and Fantasy* (Cambridge, Mass.: Harvard University Press, 2006); Keir Waddington, *The Bovine Scourge: Meat, Tuberculosis and Public Health, 1850–1914* (Woodbridge: Boydell Press, 2006); Alan L. Olmstead and Paul W. Rhode, 'An Impossible Undertaking: The Eradication of Bovine Tuberculosis in the United States', *Journal of Economic History* 64, no. 3 (2004): 734–72.
37 Most research on the history of foot-and-mouth disease focuses on the British context. See especially: Abigail Woods, 'The Construction of an Animal Plague: Foot-and-Mouth Disease in Nineteenth-Century Britain', *Social History of Medicine* 17, no. 1 (2004): 23–39; Woods, *A Manufactured Plague*.
38 Daniel Gilfoyle, 'Veterinary Research and the African Rinderpest Epizootic: The Cape Colony, 1896–1898', *Journal of Southern African Studies* 29, no. 1 (2003): 133–54; Daniel Gilfoyle, 'Anthrax in South Africa: Economics, Experiment and the Mass Vaccination of Animals, c. 1910–1945', *Medical History* 50, no. 4 (2006): 465–90; Karen Brown, 'Tropical Medicine and Animal Diseases: Onderstepoort and the Development of Veterinary Science in South Africa, 1908–1950', *Journal of Southern African Studies* 31, no. 3 (2005): 513–29; Giorgio Miescher, *Namibia's Red Line: The History of a Veterinary and Settlement Border* (New York: Palgrave Macmillan, 2012); Karen Brown and Daniel Gilfoyle, eds., *Healing the Herds. Disease, Livestock Economies, and the Globalization of Veterinary Medicine* (Athens, Ohio: Ohio University Press, 2010); Saurabh Mishra, *Beastly Encounters of the Raj: Livelihoods, Livestock and Veterinary Health in India, 1790–1920* (Manchester: Manchester University Press, 2016); Klemens Wedekind, *Impfe und herrsche: Veterinärmedizinisches Wissen und Herrschaft im kolonialen Namibia 1887–1929* (Göttingen: Vandenhoeck & Ruprecht, 2021).
39 Harrison, *Contagion*, 211–32.
40 Stephan Häsler, 'Die Veterinärabkommen des Völkerbundes von 1935: Eine Vision des Veterinärraumes Europa', *Schweizer Archiv für Tierheilkunde* 160, no. 1 (2018): 37–42; Ursula Reis, *Tierseuchen und internationaler Handel. Veterinärexperten im Völkerbund 1927–1935* (Universität Zürich: Lizentiatsarbeit, unpublished, 2005); Cornelia Knab, 'Infectious Rats and Dangerous Cows: Transnational Perspectives on Animal Diseases in the First Half of the Twentieth Century', *Contemporary European History* 20, no. 3 (2011): 281–306.

41 Zylberman, 'Making Food Safety an Issue'; Dorothee Brantz, 'How Parasites Make History: On Pork and People in Nineteenth-Century Germany and the United States', *Bulletin of the GHI Washington*, no. 36 (2005): 69–79; Justin Kastner, 'Food and Agriculture Security: An Historical Illustration of Contemporary Challenges', in *Food and Agriculture Security: An Historical, Multidisciplinary Approach* (Santa Barbara: Praeger, 2011), 5–32; Uwe Spiekermann, 'Dangerous Meat? German-American Quarrels over Pork and Beef', *Bulletin of the GHI Washington* 46 (Spring 2010): 93–110.

42 For the terms 'international' and 'internationalism': Herren, *Hintertüren zur Macht*, 13–7.

43 Madeleine Herren, 'International Organizations, 1865–1945', in *The Oxford Handbook of International Organizations*, ed. Jacob Katz Cogan, Ian Hurd, and Ian Johnstone (Oxford: Oxford University Press, 2016), 91–3.

44 Madeleine Herren, 'Introduction: Towards a Global History of International Organization', in *Networking the International System: Global Histories of International Organizations*, ed. Madeleine Herren (Heidelberg: Springer, 2014), 2.

45 Herren, 'Introduction: Towards a Global History of International Organization', 5.

46 Herren, 'International Organizations, 1865–1945', 95–6.

47 Herren, 'International Organizations, 1865–1945', 96–8.

48 Herren, 'International Organizations, 1865–1945', 96–7.

49 Paul Weindling, ed., *International Health Organisations and Movements, 1918–1939* (Cambridge and New York: Cambridge University Press, 1995); Iris Borowy and Wolf D. Gruner, eds., *Facing Illness in Troubled Times: Health in Europe in the Interwar Years, 1918–1939* (Frankfurt am Main, New York: P. Lang, 2005); Alison Bashford, ed., *Medicine at the Border: Disease, Globalization and Security, 1850 to the Present* (Houndmills, Basingstoke: Palgrave Macmillan, 2006); Susan Gross Solomon, Lion Murard, and Patrick Zylberman, eds., *Shifting Boundaries of Public Health: Europe in the Twentieth Century* (Rochester, NY: University of Rochester Press, 2008); Iris Borowy, *Coming to Terms with World Health. The League of Nations Health Organisation, 1921–1946* (Frankfurt am Main and New York: Peter Lang, 2009).

50 See for example: Pieter H. Eijkman, *L'Internationalisme Médical* (Amsterdam: F. van Rossen, 1910); *L'Internationalisme Scientifique: Sciences Pures et Lettres* (La Haye: W. P. van Stockum, 1911).

51 See also: Knab, 'Infectious Rats and Dangerous Cows', 288. About the beginnings of scholarly research on internationalism: Herren, 'International Organizations, 1865–1945', 99.

52 Société des Nations/League of Nations, *Répertoire des Organisations Internationales/Handbook of International Organisations* (Geneva: League of Nations, 1921), 6. The series was continued until 1938.

53 The research database LONSEA presents the information on the organisations and persons listed in the *Handbook* series and their networking: LONSEA, 'Searching the Globe through the Lenses of the League of Nations', accessed June 30, 2020, http://www.lonsea.de/.

54 Hans Stümke, ed., *Bibliographie der internationalen medizinischen Kongresse und Verbände*, Bibliographie der internationalen Kongresse und Verbände 1 (Berlin: O. Harrassowitz, 1939).

55 Hans Stümke (1896–?) worked as librarian at the Prussian State Library in Berlin and since 1940 at the University Library in Greifswald; he was reported missing in the Second World War. See: Cornelia Briel and Hans Erich Bödeker, *Beschlagnahmt, erpresst, erbeutet: NS-Raubgut, Reichstauschstelle und Preußische Staatsbibliothek zwischen 1933 und 1945* (Berlin: Akademie Verlag, 2013), 379.

56 Stümke, *Bibliographie der internationalen medizinischen Kongresse und Verbände*, VII.

57 The Canadian-born U.S. American Frank G. Boudreau (1886–1970) had worked from 1925 to 1937 in the League of Nations Health Organisation. Boudreau's

pamphlet *Ancient Diseases—Modern Defences* (1939) gives an overview on inter-war concepts of international public health, especially at the League, from the perspective of a participant in contemporary international health work. Frank G. Boudreau, *Ancient Diseases – Modern Defences. The Work of the Health Organization of the League of Nations* (New York: Columbia University Press, 1939).

58 Frank G. Boudreau, 'International Coöperation in Hygiene', *American Journal of Public Health* 27, no. 11 (1937): 1102–3.

59 For example: Gross Solomon, Murard, and Zylberman, *Shifting Boundaries of Public Health*; Bashford, *Medicine at the Border*.

60 Bashford, '"The Age of Universal Contagion": History, Disease, and Globalization', 4.

61 Bashford, '"The Age of Universal Contagion": History, Disease, and Globalization', 7 (citation), 13–4.

62 Zylberman, 'Civilizing the State', 22–4.

63 Anthony Cooper and Søren Tinning, 'Introduction: Thinking Theoretically and Philosophically about the Concept of the Border', in *Debating and Defining Borders: Philosophical and Theoretical Perspectives*, ed. Anthony Cooper and Søren Tinning (Abingdon, New York: Routledge, 2019), 1–4.

64 Sandro Mezzadra and Brett Neilson, 'Foreword', in *Debating and Defining Borders: Philosophical and Theoretical Perspectives*, ed. Anthony Cooper and Søren Tinning (Abingdon, New York: Routledge, 2019), XXI–XXII.

65 Laura Di Fiore, 'The Production of Borders in Nineteenth-Century Europe: Between Institutional Boundaries and Transnational Practices of Space', *European Review of History* 24, no. 1 (2016): 36–57.

66 Zylberman, 'Civilizing the State', 23.

67 Gottlieb Flückiger, 'Bisherige Bemühungen zur internationalen Bekämpfung tierischer Seuchen', *Schweizer Archiv für Tierheilkunde* 96, no. 1 (1954): 12.

68 About the process towards epidemiological intelligence in epidemics control with the aim to reduce quarantine, here described for the case of the International Sanitary Convention of 1926: Anne Sealey, 'Globalizing the 1926 International Sanitary Convention', *Journal of Global History* 6, no. 3 (2011): 432–4, esp. 434.

69 Jürgen Osterhammel, 'Globalizations', in *The Oxford Handbook of World History*, ed. Jerry H. Bentley (Oxford and New York: Oxford University Press, 2011), 89–104, esp. 90–1; Sebastian Conrad and Andreas Eckert, 'Globalgeschichte, Globalisierung, multiple Modernen: Zur Geschichtsschreibung der modernen Welt', in *Globalgeschichte. Theorien, Ansätze, Themen*, ed. Sebastian Conrad, Andreas Eckert, and Ulrike Freitag (Frankfurt am Main and New York: Campus, 2007), 7–49, esp. 20.

70 Osterhammel, 'Globalizations', 92.

71 Bashford, '"The Age of Universal Contagion": History, Disease, and Globalization', 14.

72 Mark Harrison, 'A Global Perspective Reframing the History of Health, Medicine, and Disease', *Bulletin of the History of Medicine* 89 (2015): 646–7, 661.

73 Sebastian Conrad, *What Is Global History?* (Princeton: Princeton University Press, 2016), 224–8; Jürgen Osterhammel, 'Global History and Historical Sociology', in *The Prospect of Global History*, ed. James Belich et al. (Oxford, New York: Oxford University Press, 2016), 38. For the concept of entangled history: Sebastian Conrad and Shalini Randeria, 'Einleitung: Geteilte Geschichten - Europa in einer postkolonialen Welt', *in Jenseits des Eurozentrismus. Postkoloniale Perspektiven in den Geschichts- und Kulturwissenschaften*, ed. Sebastian Conrad, Shalini Randeria, and Regina Römhild, 2nd ed. (Frankfurt am Main: Campus, 2013), 32–70. For an overview on the historiography and an analytical conceptualisation of connections in global history: Roland Wenzlhuemer, 'Connections in Global History', *Comparativ* 29, no. 2 (2019): 106–21.

74 Conrad, *What Is Global History?* 66.
75 H. R. G. Greaves, *The League Committees and World Order: A Study of the Permanent Expert Committees of the League of Nations as an Instrument of International Government* (Oxford: Oxford University Press, 1931), 11. The political scientist and economist H. R. G. [Harold Richard Goring] Greaves (1907–1984) worked since 1930 at the London School of Economics and Political Science. For a biographical note: London School of Economics, Library, 'Record: Greaves, Harold Richard Goring', accessed March 19, 2021, http://archives.lse.ac.uk/Record.aspx?src=CalmView.Catalog&id=COLL+MISC+0462.
76 Greaves, *The League Committees and World Order*, 11.
77 About experts in contexts of internationalism: Davide Rodogno, Bernhard Struck, and Jakob Vogel, eds., *Shaping the Transnational Sphere. Experts, Networks and Issues from the 1840s to the 1930s* (New York, Oxford: Berghahn Books, 2015); Helmuth Trischler and Martin Kohlrausch, *Building Europe on Expertise: Innovators, Organizers, Networkers* (Basingstoke: Palgrave Macmillan, 2014); Wolfram Kaiser and J. W. Schot, *Making Europe: Experts, Cartels, and International Organizations* (Basingstoke: Palgrave Macmillan, 2014); Daniel Laqua, ed., *Internationalism Reconfigured: Transnational Ideas and Movements between the World Wars* (London and New York: Tauris, 2011); Katharina Rietzler, 'Experts for Peace', in *Internationalism Reconfigured. Transnational Ideas and Movements between the World Wars*, ed. Daniel Laqua (London and New York: Tauris, 2011), 45–65; Waqar Zaidi, 'Liberal Internationalist Approaches to Science and Technology in Interwar Britain and the United States', in *Internationalism Reconfigured. Transnational Ideas and Movements between the World Wars*, ed. Daniel Laqua (London and New York: Tauris, 2011), 17–43; Martin Kohlrausch, Katrin Steffen, and Stefan Wiederkehr, 'Expert Cultures in Central Eastern Europe', in *Expert Cultures in Central Eastern Europe. The Internationalization of Knowledge and the Transformation of Nation States since World War I*, ed. Martin Kohlrausch, Katrin Steffen, and Stefan Wiederkehr (Osnabrück: Fibre, 2010), 9–30; Martin Kohlrausch, 'Technologische Innovation und transnationale Netzwerke. Europa zwischen den Weltkriegen/Technological Innovation and Transnational Networks: Europe between the Wars', *Journal of Modern European History* 6, no. 2 (2008): 181–95.
78 Patricia Clavin, *Securing the World Economy: The Reinvention of the League of Nations, 1920–1946* (Oxford: Oxford University Press, 2013), 7.
79 Peter M. Haas, 'Introduction: Epistemic Communities and International Policy Coordination', *International Organization* 46, no. 1 (1991): 1–35.
80 Haas, 'Introduction', 3.
81 Haas, 'Introduction', 3–4.
82 See also: Davide Rodogno, Bernhard Struck, and Jakob Vogel, 'Introduction', in *Shaping the Transnational Sphere. Experts, Networks and Issues from the 1840s to the 1930s* (New York, Oxford: Berghahn Books, 2015), 12.
83 Greaves, *The League Committees and World Order*, 11.
84 See also: Rodogno, Struck, and Vogel, 'Introduction', 2.
85 For an overview: Jan Ruzicka and Vincent Charles Keating, 'Going Global: Trust Research and International Relations', *Journal of Trust Research* 5, no. 1 (2015): 13–5.
86 Ruzicka and Keating, 'Going Global', 15–21; Ute Frevert, *Vertrauensfragen: Eine Obsession der Moderne* (München: Beck, 2013), 15–6.
87 Frevert, *Vertrauensfragen* (translation: CK).
88 Frevert, *Vertrauensfragen*, 43, 147–208. For further discussion and approaches towards trust in historical research: Jörg Baberowski, ed., *Was ist Vertrauen? Ein interdisziplinäres Gespräch* (Frankfurt am Main, New York: Campus Verlag, 2014).
89 Geoffrey A. Hosking, *Trust: A History* (Oxford, New York: Oxford University Press, 2014), 4.
90 Hosking, *Trust*, 45–6.

2 Animal diseases and the global 19th century

> In no age before ours has the sanitary control of animals been of equal interest to that which it presents today, because never has the need to multiply and conserve them been so obvious.... The international relations which the railways have created, the new markets which freedom of trade has opened up for the importation of foreign animals, have changed the sanitary conditions of the various states of Western Europe.... It can be seen that the seemingly humble questions of animal health regulation are transformed, when considered from this point of view, into a great question of commerce, industry, and consequently of wealth and civilisation.[1]

In the foreword to his book on state veterinary organisation published in 1873, Jean Reynal, then director of the French veterinary school of Alfort, tried to draw attention to what he considered a new situation of animal disease problems in Europe with worldwide relevance.[2] He outlined that in the past, epizootics had frequently appeared in massive dimensions in Europe, but they had done so intermittently due to geographical distances. By now, he argued, there was a more frequent influx of diseases into Europe from both near and faraway regions as a consequence of the rapidly growing net of worldwide commercial interconnections. This situation, he explained, resulted in a variety of complex interrelated problem areas and produced new challenges of epizootics control.[3] Reynal's analysis of changes in the mid-19th century situation concerning the spread of animal diseases reflected the worries of many contemporary experts. The entangled and multifaceted processes of change during the long 19th century, which Jürgen Osterhammel has called "the transformation of the world",[4] influenced people's lives and living environments on a global scale, and these processes also profoundly affected the understanding and handling of animal diseases.

This perception of a changing world, with a rapidly increasing densification of transport and travel connections, was one of the main motivations for veterinary and scientific experts to initiate activities of international cooperation to deal with the spread of animal diseases. Three main, interconnected developments had a decisive influence on initiatives of international veterinary cooperation in the late 19th and early 20th centuries. First, the process of integrating animal disease problems into the responsibilities

DOI: 10.4324/9781003033998-2

of the state administration created new structural conditions to deal with on animal diseases. The development of state-funded institutions for the education of veterinary professionals and of state veterinary services also facilitated international networking and cooperation efforts. Second, the technical innovations and the pathbreaking developments in medical and microbiological research of the late 19th and early 20th centuries had a profound impact on the approaches towards animal diseases of the time by creating new possibilities for their prevention, control and treatment. Third, the increase in globalised trade relations and processes of industrialisation in the livestock and food production sector transformed the contexts and conditions of animal farming and contributed to creating new global spaces and pathways of disease transmission. These processes of transformation influenced and increased the tasks and responsibilities of veterinary experts and brought along new questions and challenges that required coordination efforts beyond national borders. During the 19th century, therefore, the agenda in the field of animal disease control expanded continuously, which contributed to the legitimisation and consolidation of the still young academic discipline of veterinary medicine.

The state and the veterinary expert

At the time when the first international initiatives addressing animal diseases began in the mid-19th century, rinderpest, or cattle plague, was in English also called 'Russian plague'; in German, it was referred to as 'Orientalische Rinderpest'.[5] The geographical reference was not unusual in the naming of contagious diseases. In the case of rinderpest, it identified the 'East' from a European perspective as a site of otherness and origin of a dangerous disease imported into Europe.[6] Western or Central European veterinarian experts such as Jean Reynal saw these 'Eastern' or 'oriental' regions as starting vaguely at the borderlands of the Austro-Hungarian Empire and in Northern Africa.[7] Usually Russia was indicated as one of the main places of origin of rinderpest. Epidemics and epizootics, and the measures against them, were often described in military terms, they were 'intruders' from 'outside' and thus 'invading' clearly defined territories which needed to be 'defended'.[8] These ideas of epizootics as foreign threats coincided with concepts of nationalism and ideas of European superiority. The Western and Central European perception of epizootics such as rinderpest originating in Eastern European or Asian countries was connected with century-old historical experiences of animal diseases travelling frequently along the trans-Eurasian routes through activities of trade and in the wake of wars. As rinderpest was mostly transmitted to Austria-Hungary, for example, through the trade in cattle from Eastern Europe,[9] these routes were considered to be the main entries through which rinderpest spread to Europe. Animal diseases, however, could be transmitted in many ways and geographical directions: multiple factors related to human living conditions and the ways of interacting with animals contributed to the spread of epizootics.[10] Until the late 19th century, and in many parts of the world continuously in the 20th and 21st centuries, horses, donkeys, cattle and other animals were essential means of travel, transport and labour. As Madeleine Ferrières has described, livestock was mobile and omnipresent, domestic animals were thus in frequent and close contact with other animals as well as with humans, and this "constant tide of livestock on foot involved all species".[11] As a consequence, contagious and non-contagious animal diseases appeared frequently in both rural and urban contexts. The persons in charge of animal health—farmers, blacksmiths, peasant healers, stable masters, physicians, and since the 18th century more and more academically educated veterinary specialists—were confronted with a variety of emerging, overlapping and disappearing diseases of many species. Sometimes outbreaks of epizootics coincided with human epidemics, making a differentiated diagnosis even more challenging.[12] The uncertainties of disease origins, pathways of transmission, disease carriers and of a clear diagnosis made it a tremendous challenge to establish effective responses to prevent and control contagious diseases of animals.

In the case of epidemics control, measures frequently centred on spatial strategies, such as the isolation of sick persons and the development of sanitary barriers to protect a territory from incoming diseases. In the late Middle Ages and early modern period, such cordons sanitaires were developed in the Mediterranean region to keep out the plague and other contagious diseases. They manifested as border fortifications to protect a territory from epidemics by military means, including a detailed system of sanitary practices and surveillance measures to control and regulate the entry of humans into a territory. The control measures varied and could include obligatory health certificates for travellers, travel restrictions, and a system of communication and information.[13] In addition, the practice of quarantine imposed on arriving travellers for a certain period of time served as a preventive measure combining temporal and spatial approaches of confinement to reduce possible disease transmissions.[14] The measures against the spread of contagious animal diseases were oriented on the regulations against human diseases, but had some specific characteristics. In the course of the 18th century, the problem of epizootics gained massively in significance for many European governments because of severe and recurrent outbreaks of different contagious animal diseases. These experiences resulted in more regular and profound interventions in animal disease control by the state authorities.[15] Beginning in 1711, rinderpest spread to many European countries, introduced probably by cattle transported on the Danube.[16] Outbreaks of foot-and-mouth disease and other diseases also appeared repeatedly in Europe.[17] For decades, there were repeated waves of epizootics in many regions, and the situation often took on dramatic proportions. Livestock, however, did not only die from a highly contagious epizootic such as rinderpest, but also from the measure of culling herds ordered by state authorities. As Madeleine Ferrières has outlined, the introduction of the so-called stamping-out to eradicate an animal disease in a defined territory was, in such radical form, a novelty in 18th-century Europe—it was a rigorous intervention of the authorities into the property of livestock owners, and it reflected a parallel discourse on demonstrating the supremacy of humans over the animal world.[18] Ferrières has stressed in this context that the estimation of historians of over 200 million dead cattle caused by the animal diseases in 18th-century Europe must be viewed critically, as the sources do not always clearly distinguish between cattle killed by disease or by systematic culling.[19] The number of diseased and dead livestock was, in any case, enormous, and resulted in a difficult situation for livestock owners. The Italian Giovanni Maria Lancisi, who worked as papal physician in the late 17th and early 18th century, initiated a number of systematic measures against epizootics to protect the papal state by combining several old and innovative elements of animal disease control.[20] He introduced strict veterinary sanitary practices such as especially stamping-out even of suspected animals, the burying of animal carcasses, inspection of animal products, and control of the movements of livestock at borders by quarantines and prohibiting of transports. The epizootics regulations included severe punishments for those who did not follow.[21] Lancisi successfully managed to remove the disease from the papal territory. In the course of the 18th century, several governments in Europe followed his examples and established regulations to contain animal diseases locally and to intercept their spread.[22] Many of these interventions against epizootics were conducted with the military force of absolutist states—as Ferrières states in her description of the practices in England, Italy and France, "the victories against epizootics were military victories".[23] The policy of rigorous culling was highly controversial and had many opponents, most notably the affected livestock owners. The economic, social and political consequences of epizootics became also apparent in context of the trade with animal products. Killing a diseased animal did not necessarily mean that its meat was removed from the market; often, it was still sold.[24] As Ferrières outlines, in parts of 18th-century Switzerland, for instance, it was considered charitable to distribute ominous meat from (maybe) diseased animals to the poor.[25] The question of whether meat from infected animals should still be consumed would continue to provoke discussions also in the later international context, for example in case of tuberculous livestock. For people who were dependent on animal husbandry for their livelihood, the waves of epizootics together with the strict control measures of the state authorities often generated precarious living situations, causing food shortages, interrupted commercial cycles, limited supplies and decimated incomes, and thus resulting in famines and in social-economic misery.

The increasing approaches and interventions of state authorities in order to contain the spread of epizootics were part of the reforms and expansions of administrative structures in 18th- and 19th-century Europe. As a result of the experiences with epizootics, the task of handling the problem of contagious diseases of animals was integrated in many countries into the responsibilities of the state administration, a process which continues in many parts of the world today. One of the main concerns of the new veterinary systems was the foundation of state-funded veterinary schools for the education of veterinary specialists, the first ones in Lyon in 1762, Alfort (near Paris) in 1766, and Vienna in 1767.[26] Other European governments followed with the foundation of veterinary schools, and soon there existed a considerable network of new institutions dedicated to veterinary education and research. Their main task, at the beginning, was usually to secure the health of horses for military usage; gradually they extended the focus to diseases of those animals which were economically relevant, meaning all sorts of livestock.[27] Many European and non-European countries followed this development in the 19th and 20th centuries. Researchers have recently outlined that the establishment of veterinary schools was part of a broader and longer process of medical and scientific interest in animal health and its connection with human medicine.[28] The close connections between specialists of human and animal medicine certainly fostered the process of establishing veterinary medicine as a new academic discipline with a regulated system of education integrated in scientific research, constructing a distinction of the new veterinary experts from the traditional 'lay' healers.[29] This academic-scientific professionalisation of veterinary medicine and its implementation in practice as the main valid approach to address animal diseases was not without conflicts. The new institutions and the associated regular funding by the state created opportunities to advance research and experiments and contributed to further differentiate and enlarge the academic discipline of veterinary medicine. In this context, the French term *épizootie* was introduced in 1765 by the veterinarian Claude Bourgelat, the founder of the Lyon and Alfort veterinary schools. The term was based on the notion of epidemic to indicate diseases affecting great numbers of animals.[30] The European experiences with waves of epizootics and the development of new concepts, practices and institutions to implement systems of animal disease control in Europe since the 18th century had significant consequences for later international efforts. Many of the elements of today's state-organised veterinary sanitary responses to epizootics were developed or elaborated during the 18th and 19th centuries—the establishment of veterinary authorities funded by the state and integrated in its administrative structures, the further development of legislation and regulatory mechanisms, and the policy of strict culling of sick animals (or suspected groups of animals). As another consequence of the processes of veterinary institutionalisation, professional differentiation, and integration into the tasks of the state administration, scientific research and literature on the pathology of animals and on the organisation of veterinary control efforts increased considerably since the late 18th century. The new state-funded veterinary schools symbolised the regulatory role of the state in questions of animal health and veterinary education. They enabled the state administrations to create permanent veterinary services run by experts, equipped with specialised veterinary knowledge, who regularly supervised and analysed the territorial disease situation and who were charged with developing and implementing strategies against epizootics. The veterinary schools also facilitated scientific exchanges among the newly created members of a veterinary scientific community. However, there existed a broad variety of different veterinary practices, scientific traditions as well as administrative and regulatory structures in the countries that traded with animals, especially on a global scale. Furthermore, despite the establishment of sanitary veterinary borders with their range of anti-disease practices, especially the quarantine measures, the mobility of animal diseases across borders remained a core difficulty and a key challenge for the upcoming international projects of animal disease control.

Animals and their diseases under the microscope

For centuries, animals and the research into their anatomy and their diseases played an important role in the production of scientific knowledge and human medicine. Since the middle of the 19th century, the new research approaches in microbiology, the scientific debates about contagiousness, and the identifying of pathogens and their characteristics profoundly transformed also the field of veterinary medicine.[31] Scientists provided through laboratory work more and more detailed analyses of disease pathogens, the pathways of disease transmission and the vector animals, and they developed new options for the prevention, control and treatment of diseases. The pace of scientific research in veterinary and human medicine accelerated with pathbreaking research findings. Robert Koch published a description of the anthrax bacterium in 1876, and of the tuberculosis pathogen in 1882.[32] In the 1880s, Louis Pasteur presented methods of vaccination against anthrax and rabies.[33] In 1893, Theobald Smith and his colleague F. L. Kilbourne, working in the Bureau of Animal Industry (BAI) of the U.S. Department of Agriculture, showed that Texas fever, a parasitic cattle disease, could be transmitted by a vector animal—a finding that opened up new directions for research into numerous vector-borne diseases of animals and humans.[34] Furthermore, the field of virology gained in importance. Friedrich Löffler and Paul Frosch analysed the viral pathogen of foot-and-mouth disease in 1898.[35] In 1902, Maurice Nicolle and Adil Mustafa published a description of the rinderpest virus.[36] The interconnection of research in the areas of human and veterinary medicine played a central role for the development of new scientific approaches. The possibilities of laboratory research provided researchers with scientific opportunities to develop tools complementing the measures of sanitary veterinary control, as well as with additional scientific legitimation and esteem.[37] Animals took on various roles in these developments. They were objects of research for the understanding of diseases, prevention and therapies, also because of the links between the human and the animal forms of zoonotic diseases. Animals were also needed as an essential material resource for researchers, as living objects in the laboratory for experiments on the development of medical treatments, diagnostic tools and vaccination trials, as well as for the production of biological products.[38] This resulted in the need for huge numbers of standardised animals, for example rabbits and laboratory mice.[39] Laboratory-based research thus is until today unthinkable without the large-scale usage of many different species of animals. From the beginning, the extensive use of animals in laboratory research also caused controversies: the opponents of animal experiments formed one of the early organised transnational civil society movements in the late 19th century.[40]

Ideas about the close connection between a number of human and animal diseases and their transmissibility between species had existed since antique times.[41] In 18th- and 19th-century Europe, fears of diseases transmissible from animals to humans were widespread, especially in the rural world.[42] Researchers debated the transmissibility of certain diseases from animals to man, especially rabies, anthrax, scabies (a parasitic skin disease) and glanders (a bacterial disease of horses and donkeys), increasingly since the 17th century.[43] In 1837, the French physician Pierre François Olive Rayer proved that glanders was transmissible to other animal species and also to humans. The disease was especially relevant because of the significance of horses in transport, travel and warfare.[44] As can be seen by the increase in medical literature on the question of disease transmission between different animal species including humans, the topic area linking human and veterinary medicine, since the mid-20th century called veterinary public health, moved more and more into the focus of research since the middle of the 19th century.[45] The term *Zoonosen* was used by the German physician Rudolf Virchow in 1855 to indicate "infections through contagious animal poisons".[46] In 1863, the Munich veterinary professor Wilhelm Probstmayr defined a double meaning of the term, firstly as animal diseases and secondly as human diseases transmitted by an animal.[47] The Italian researcher Bruno Galli-Valerio, who worked at the University of Lausanne, published in 1894 a study on zoonoses which he described as transmissible diseases from animals to man.[48] In 1958 the Joint WHO/FAO Expert Committee on Zoonoses provided a definition as "those diseases and infections which are naturally transmitted between vertebrate animals and man"[49] without making a difference in the direction of the transmission. This general definition is still in use today, although there have been further attempts at specification.[50]

Linked to the studies of pathogens and zoonotic disease relationships in the 19th century were research efforts on vaccination as a means of prevention of epidemics and epizootics. The vaccination of human beings was a public health instrument to provide protection both for the individual and for the collective by aiming for so-called herd immunity.[51] The term, clearly derived from the area of animal health, appeared first in 1916 in veterinary contexts in the United States and was used as a concept in human medicine since the early 1920s.[52] In the case of the development of vaccines for animals, the motivation was, in most cases, protecting herds for economic reasons.[53] Traditional methods of obtaining immunisation, mainly from the Ottoman Empire and the Middle East, Asia and Africa, increasingly came into the focus of European researchers since the 17th century, for example the ideas on smallpox inoculation.[54] In the field of epizootics control, methods of immunisation to prevent sheep pox were practiced in some Eastern parts of Europe already at least since the 15th century; in France, a scientific debate about sheep pox immunisation started in the mid-18th century.[55] In the case of rinderpest, numerous experiments on obtaining immunisation were carried out in several countries during the 18th and 19th centuries. The procedures were risky and the outcomes not at all satisfying, since they resulted not only in the loss of large numbers of animals, but also entailed the possibility of spreading the disease further through the experiments.[56] For the development of safe, effective and easily available methods of immunisation, the successes against human smallpox in the late 18th century seemed promising. Material from cowpox was from then on used for vaccination against smallpox in humans, taking advantage of the disease relatedness across the species, even though the background remained unclear for a long time.[57] Also in the case of rinderpest, attempts to vaccinate cattle with cowpox were undertaken during the 19th century because the disease symptoms appeared to be similar; unsurprisingly, the results were a failure.[58]

In the context of the microbiological research of the late 19th and early 20th centuries, scientists focused intensely on the development of new vaccines for both human and animal diseases. The research on vaccines for humans and animals was always closely linked in interdisciplinary scientific work and needed the expertise of veterinarians—the vaccination of animals served as “a model, a tutor and a complement”[59] to the development of vaccination for humans. Pasteur and his team in Paris focused, among other diseases, on anthrax, chicken cholera and rabies.[60] Anthrax was feared as a disease of both livestock and humans in rural and industrial contexts, since it could be transmitted through the pathogens in the soil but also via animal products such as wool, hides and skins.[61] In 1881, Pasteur conducted a famous public experiment with a herd of sheep, demonstrating the efficacy of his newly developed vaccine with an attenuated pathogen to prevent anthrax in animals.[62] A few years later, the development of a vaccine against rabies became a further celebrated success of research. Rabies, a zoonotic disease transmitted from animals to humans or to other animals, was enormously feared, even though being comparatively rare. As Kathleen Kete has shown for 19th-century France, rabies represented the emergence of the “*bête humaine*” because of its atrocious symptoms and its certain lethality: it appeared as the darker side of 19th-century discourses on human distinction towards the animal world, a confusing challenge to narratives about human progress.[63] The widespread fear of the disease contributed to the global media hype that followed Pasteur’s successful experiment of vaccinating a child in 1885.[64] The Pasteur vaccine was also used for the immunisation of animals against rabies.[65] Further developments in the field of vaccination followed in the next decades and again showed the close interdependence of veterinary and human medicine—successes in animal vaccination research were linked to efforts in human medicine to improve public health, and animal resources and experimentation remained in any case the basis of all efforts to produce new vaccines. In 1886, for example, the U.S.-American researchers David Salmon and Theobald Smith at the BAI presented successfully the method of inactivating pathogens by treatment with heat for the production of vaccines, in the case of a salmonellosis.[66] Researchers in the early 20th century further developed vaccine improvements and combined the principles of attenuation and inactivation of pathogens through heat and chemicals with the use of adjuvants.[67] In 1924, the French veterinarian Gaston Ramon, who later served from 1948 to 1959 as director of the OIE, presented a vaccination against tetanus with a toxoid by treating the toxin of the pathogen with heat and formaldehyde while keeping the immunological effects.[68] These efforts enabled further research and the production of vaccines against a number of human and animal diseases. In the case of foot-and-mouth disease, the development of vaccines made progress in the 1920s and 1930s, and since the late 1930s research institutions, at first the Riems institute in Germany, started with efforts of an industrial production of vaccines for the large scale immunisation of herds.[69] In the case of rinderpest, the first effective vaccines produced since the early 20th century were helpful to contain the disease but often did not produce long periods of immunisation. Since the 1920s a so-called caprinised vaccine was widely in use (creating a reduced virulence of the pathogen through the serial passage in goats).[70] A new vaccine version produced by Walter Plowright and his team in Kenya in the 1950s, improved in the later decades, made possible the large-scale global vaccination campaign against the disease in the second half of the 20th century.[71]

The development and production of safe, effective and transportable vaccines and other biological products for diagnostics and treatment was expensive, sometimes dangerous, and needed specialised research institutions. It also required a framework of legal regulation, both at the national and the international level, for example for the registration of new biological products.[72] Further, the process of vaccine production and the organisation of large-scale vaccination campaigns needed financial resources. For both the national and the international level of epizootics control, the research concerning vaccination, diagnostic tools and treatment offered new options. In the context of these developments, the academic discipline of veterinary medicine underwent profound transformations.[73] The responsibilities of veterinarians, the conceptual-scientific basis of their profession and education, and the methods and scientific approaches changed and enlarged considerably. Many practicing veterinarians expressed scepticism at the new research foci that seemed to sideline the multiple sources and contexts of the emergence and spread of epizootics.[74] For the control of epizootics, the previously devised veterinary sanitary instruments remained absolutely central.[75] But the development of vaccination for a number of diseases became the key for organising large-scale prevention campaigns within and beyond national borders, paving the way for the global eradication of animal diseases such as rinderpest that were suitable for global vaccination programmes. The boost in microbiological research concerning the nature, spread, prevention, diagnosis and treatment of both human and animal diseases was also reflected in the increase of specialised scientific publications. The new research efforts and their implementation in practice revealed the different situation of national or colonial veterinary services worldwide, many of which could not afford the professional staff, the financial means or the scientific equipment to benefit in a permanent way from the new instruments developed by bacteriological and virological research. The new developments created pressure on state authorities to improve their efforts of monitoring, documentation and information concerning the national, regional and local situation of animal diseases, and to publish their results and disease statistics. By 1900, the documentation of the progress in the control of animal diseases and of scientific achievements in an official bulletin was also a question of national scientific prestige. The regular evaluation of disease situations invited for the comparison between the different national efforts concerning the control of animal diseases within a territory. Governments used sanitary arguments to legitimise their public health and economic policies, mainly with regard to interventions in the trade in animals and to demand activities from other governments concerning animal disease problems. The innovative scientific discoveries and developments were therefore important at both the scientific-medical and the political-economic level. They opened up numerous new perspectives for the control of animal diseases, while their implementation also revealed and underscored underlying power asymmetries and inequalities on a global scale.

Animal industries and the globalisation of trade

In the context of the processes of industrialisation, colonisation and globalisation in the middle and late 19th century, a number of intertwined political, economic and social factors contributed to the increasing perception of animal diseases as globally relevant problems. New means of travel and transport such as railways and steamships connected the world at a greater speed, and business companies extended their global supply chains. The new transport networks also enabled the faster spread of contagious animal diseases over long distances.[76] Conventional veterinary sanitary control measures at borders, such as quarantines, were disruptive factors for the globalised movements of trading goods. In order to meet the growing demand for livestock and animal products resulting from the increase of the world's population, the production of livestock in many countries grew rapidly.[77] In Argentina, one of the new meat exporters for the world market, the number of cattle doubled between 1875 and 1908.[78] The increasing industrialisation in many countries additionally created a rising demand not only in meat and milk but also in animal products such as fats, hair, bones, wool and leather. New forms of intensive livestock farming processed animals in previously unimagined numbers, technical dimensions and speed. The industrialisation of the 19th century thus gradually but profoundly transformed the economic structures, conditions and practices of animal farming and trading. With increasing processes of specialisation of production and economic monopolisation, a greater amount of animal products in shorter periods of time and at lower costs could be produced.[79] At the same time, the activities of colonisation contributed to deforestation and the transformation of land for agricultural usage, pastures and settlements, and to the creation of new contact zones for the transmission of pathogens among animals and from animals to humans. Both colonisation and warfare activities needed resources and supplies in form of horses and livestock. The combination of these factors contributed to a new dimension of animals and animal products being transported in huge numbers as economic resources over large distances.

From the middle of the 19th century, the news of epizootics outbreaks circulated around the world quicker than before due to new communication technologies, and news about the emergences and spread of animal diseases seemed to appear frequently. From the 1830s and 1840s onwards, outbreaks of bovine pleuropneumonia, a contagious bacterial disease of cattle affecting the lungs, as well as of foot-and-mouth disease, appeared repeatedly in Western European countries.[80] In 1865, as a result of the opening of markets due to free-trade policies, rinderpest was introduced into Britain by a shipment of cattle from Russia. The outbreaks of the disease in 1865/66 devastated the British cattle sector and caused a public outcry about the insufficient measures of the authorities. Following this outbreak, a systematic veterinary legislation with regular veterinary inspectors was established.[81] Continental European countries were affected as well by the rinderpest outbreaks during those years; the disease also appeared on occasion of the war of 1870–71 between France and Germany.[82] In Asia, epizootics from Eastern China spread to Japan after the opening of the country during the Meiji period.[83] Often, animal diseases travelled along the trade routes. In South Asia, epizootics of livestock such as rinderpest and foot-and-mouth disease appeared frequently in the second half of the 19th century, in some regions being enzootic.[84] In North America, the livestock sector suffered for decades from outbreaks of Texas fever, a disease of cattle with a high mortality rate, transmitted through a parasitic pathogen spread by ticks.[85] A catastrophe on multiple levels was the rinderpest panzootic in sub-Saharan Africa from the late 1880s until the early 20th century. In most parts of sub-Saharan Africa, the disease had been rare or practically non-existent before. The accounts of the new outbreaks traced the beginning to the Italian army and its cattle imports from India into Ethiopia in 1887.[86] The spread of rinderpest had immense social, economic, environmental and political consequences. For years, the disease caused massive famines and economic disaster among the population, resulting in innumerable human deaths. Rinderpest not only decimated livestock but also killed wild animals such as zebras and antelopes in large numbers, and thereby caused significant transformations in the African ecological systems.[87]

Veterinary experts in Europe gradually recognised the need to focus far more beyond the national and local problems of animal diseases and to draw attention to the role of global interconnectedness in the spread of diseases. In his 1875 *Manual of Veterinary Sanitary Science and Police*, the British veterinary surgeon George Fleming argued to address this global dimension with a more comprehensive version of a veterinary sanitary organisation.[88] Fleming had made a career as a veterinarian in the British military in colonial contexts and was a proponent of the germ theories of diseases and of veterinary public health in Britain.[89] His book also took up a core dilemma of veterinary work: a functioning veterinary administration needed to tackle the mobility of animal diseases in its links to in trade, travel, animal-farming and food production, while most effective veterinary sanitary measures presented hindrances to the flow of commerce. This was a problem in context of the British free trade policies, as well as with regard to imperial trading networks. Fleming suggested to organise a network of reformed national veterinary sanitary services with a distinct orientation towards concerted solutions:

> With a view to render sanitary measures effective, in this respect, the regulations with regard to the prevention of contagious diseases should be international, and as nearly uniform as may be possible.... An inspection of the animals should be made by thoroughly trustworthy veterinary officials, whose certificate—specifying the number and species of the animals, date of inspection, their freedom from actual disease, and the absence of suspicion with regard to contamination—should accompany them.... Again, a country in which a contagious disease has suddenly appeared should be bound to report the occurrence to other countries which receive its exported animals.... Until these international measures are introduced, the countries into which animals are exported will run great risk of infection among their own herds and flocks with diseases carried by the new arrivals.[90]

Fleming's assessment thus highlighted that the veterinary services of trading countries needed to be modernised, improved and adjusted to uniform standards; only then would measures such as health certificates for traded livestock make sense. Indeed, the patchwork of different standards and regulations of veterinary control contributed to difficulties in the international trade in livestock and animal products. Fleming's outline of a network of official veterinary sanitary services and of uniformity in the control systems was a much-discussed issue among veterinarian experts in the second part of the 19th century. The problem became particularly relevant due to the increase in the global trading volume in animals and animal products. Livestock had always been considered as economic goods that needed to be optimised in their quality and maximised in their numbers and individual size, if possible, by improvements in breeding, nutrition and medication. These practices intensified again during the 19th and 20th centuries. The demand for greater economic profitability led to changes in the bodies and behaviour of industrially raised animals, with the aim to produce livestock that were easier to process and that generated higher economic profits.[91] The farming, slaughtering and processing of animals was more and more being removed from the sight of consumers, at least of those living in urban contexts in industrialised countries. Many cities—mainly in Europe and North America, but also in colonial contexts—built large slaughterhouses on their outskirts with the intention of centralising the slaughter of livestock and banishing it from the city centres.[92] Famous examples included the Parisian slaughterhouse system built in 1867 at La Villette or the Chicago Union Stock Yards, which started its production in 1865. The new and large animal factories raised and processed huge numbers of livestock for a globalised mass market.[93] The terrible working conditions for employees in the meat-packing industry in Chicago were famously described by Upton Sinclair in his novel *The Jungle* (1905), which caused a public scandal in the United States.[94] The increasing profitability of livestock production systems was therefore closely linked to exploitation and appalling working and hygiene conditions in factory farming.[95] The new slaughterhouse systems were part of an increasingly globalised meat market.[96] A study by the U.S. Department of Commerce from 1925 indicated the amount of around 3 billion pounds of exported meat (average in total) as the annual shipment before the First World War from non-European countries (mostly the U.S.A., Argentina, Uruguay, and to a lesser extent Australia and New Zealand) to Europe.[97] Through improved techniques in food preservation and refrigeration, companies could transport enormous amounts of processed meat across long distances. At the same time, in most countries, traditional forms of animal husbandry continued or coexisted alongside new industrial livestock farming. The use of animals for transport and work, especially in rural contexts, continues in many areas of the world in the 21st century.

The industrialisation of the livestock sector required a professional workforce and new logistics for adequate systems of hygiene and veterinary control in animal factories and slaughterhouses. The focus of veterinary medicine on diseases of animals and the handling of epizootics was enlarged to include more tasks related to veterinary public health in order to secure food hygiene and human health.[98] This process gained momentum when veterinary experts, governmental authorities, producers and consumers began to worry about a number of meat- and milk-borne human diseases with pathogens identified and scientifically described by researchers in the second part of the 19th century. One of the central issues of human and animal health that came to the fore was the problem of trichina worms in hogs. The discovery of the parasites, which can cause the disease trichinosis in humans, in more and more hogs raised questions about the adequacy of slaughter controls and the quality of food inspection in exporting countries. In terms of veterinary organisation, the problem of trichinosis and other meat-borne diseases influenced the development of professional meat inspection.[99] Trichinosis appeared often in countries with high pork consumption such as Germany, where outbreaks during the 1860s caused public outrage.[100] As a result, several national health authorities started to introduce regular meat inspection procedures by professional meat inspectors, who extended their work to also check for signs of animal tuberculosis and other diseases.[101] The flow of scientific research publications, statistical analyses, case study descriptions and meat inspection manuals in the late 19th and early 20th centuries reflected the enormous interest of experts, state authorities and the public in the improvement of meat safety. The *Handbuch der Fleischbeschau* (Handbook of Meat Inspection) by the Berlin veterinary professor Robert von Ostertag in 1892 became, for example, one of the international standard handbooks in the field.[102] Ostertag was one of the main organisers of a scientifically based system of meat inspection and food hygiene in Germany. His comprehensive bibliography of meat inspection, published in 1905 as a supplement to his *Handbuch*, clearly shows the topicality of the issue through the large number of worldwide scientific literature listed.[103] The issue of trichinae parasites became particularly relevant for governmental administrations because the disease was used as an argument for establishing non-tariff trade barriers. The doubts—real or purported—about the safety of meat products from certain countries were a major problem for the meat business in exporting countries. From the 1880s, the problem of potentially unsafe American meat products imported into Europe became a prominent example of sanitary-related commercial conflicts as a number of governments, among them France and Germany, banned meat imports.[104] The problem engendered heated political debates and press coverages on both sides of the Atlantic on the relation between sanitary considerations and protectionist policies. The case illustrated the multi-faceted connections between the cultural habits of meat consumption, the influence and visibility of medical-sanitary expert knowledge, the entangled transatlantic economic relations, the conflict between interests of free trade and of protectionism, and the use of nationalist sentiments in public health and foreign policies.[105] It also revealed the necessity of finding international agreements and common regulations to prevent the disease and avoid future trade disputes.[106]

The trichina parasite was only one of numerous disease problems which came to the fore as a prominent and globally relevant issue of animal health in the late 19th and early 20th centuries. An informative overview of the inclusion of veterinary matters into debates of public health was available in the German journal *Deutsche Vierteljahrsschrift für öffentliche Gesundheitspflege*, which appeared from 1869 to 1915.[107] The journal showed how issues of animal diseases were part of a multiplicity of connected topic areas and contexts—food hygiene, sanitation in trade-related questions, infectious diseases and their treatment, industrial and agricultural labour issues, questions of waste disposal, and urban sanitary planning. It included in its regular overviews on the sanitary situation both epizootics and diseases with a zoonotic potential. Anthrax, for instance, previously a disease occurring mainly in rural areas, by now was discussed also as a disease affecting industrial workers in contact with animal wools and skins in textile factories and tanneries.[108] In particular, the connection between animal and human tuberculosis was in the late 19th and early 20th centuries one of the important questions of public health and veterinary sciences.[109] Animal tuberculosis, especially its bovine form, and its possible transmission to humans was a highly disputed topic which was linked to the security of food of animal origin, the technical-scientific possibilities to protect against disease transmission, and the question of state responsibilities in public health.[110] At its core was the problem that the scientific insights on the potential transmissibility of tuberculosis between different species had resulted in multiple new uncertainties about the infected animals and a possible disease transmission to humans via milk and other animal products.[111] The new testing with tuberculin, used as a diagnostic, indicated high numbers of infected animals. In Germany, for example, Ostertag estimated in the 1890s that about a quarter of all cattle were infected with tuberculosis pathogens.[112] For public health authorities, veterinarians, and the food-producing industry, the numbers and the complex entangled uncertainties of the disease transmission were a predicament.[113] The discussion about bovine tuberculosis resulted in increased attention to food- and especially milk-borne diseases in general, another field of veterinary engagement in public health which gained momentum. Regularly, there were doubts on the safety of milk, for example in 1908 when U.S.-American bacteriologists counted several thousands of microbes in one drop of raw milk (depending on the time after milking).[114] Among the newly described milk-borne diseases were infections such as brucellosis, a bacterial disease appearing in different animal species and being transmissible to humans, which from this time on regularly appeared on the agenda of international meetings in the 20th century.[115] The procedure of pasteurisation with a procedure of heating to eradicate pathogens seemed a promising way out of the milk dilemma. Since the 1900s, first states began to introduce legislation for the obligatory pasteurisation of milk.[116] However, since many experts assumed that it would decrease the nutrients in milk, and the scientific controversy about the transmission of the bovine pathogen of tuberculosis continued, the procedure of pasteurisation remained a controversial instrument.[117]

Veterinary experts were thus confronted in the late 19th and first half of the 20th century with a spectrum of animal disease challenges that urgently required new responses and structures beyond national borders. The global spread of contagious animal diseases was not in itself a new phenomenon, but the intensified transport of animals and the increased consumption and use of animal products influenced disease outbreaks. Many veterinary and scientific experts warned about the globalised dangers of contagious diseases of animals, and tried to draw attention to the new needs of animal and public health that appeared in context of mass industrial farming. The findings of scientific and biomedical research during these decades contributed to a new understanding of the origins, characteristics and spread of animal diseases and revealed unknown complexities, creating new uncertainties as well as new opportunities of disease control. Since the middle of the 19th century, veterinary and public health experts therefore faced an increasingly complex, on many levels globally entangled, situation of animal diseases. The working areas and responsibilities of veterinary and sanitary services expanded to include, in addition to epizootics control, increasingly tasks connected to veterinary public health.[118] Sanitary measures of animal disease control applied in national settings lost none of their relevance, but their shortcomings became apparent in view of the cross-border mobility of pathogens and the complex entanglements of disease problems. Moreover, the established state veterinary services followed different administrative and scientific orientations, traditions and standards. Consequentially, there were enormous disparities between the respective national approaches towards animal disease control. At the turn to the 20th century, in many countries and colonies there was only a rudimentary or even no state-organised veterinary administration to control the emergence and spread of animal diseases. This situation was the source of scientific and political disagreements and the background for trade conflicts and sanitary disputes. The need to internationalise the agenda for animal disease control and to initiate new cooperation efforts therefore became, in the eyes of many contemporary observers, an increasingly urgent concern.

Notes

1 Jean Reynal, *Traité de la Police Sanitaire des Animaux Domestiques* (Paris: Asselin, 1873), V, VI, IX (translation: CK).

2 About Jean Reynal (1816–1893) and his position in French academia: Ronald Hubscher, *Les Maîtres des Bêtes: les Vétérinaires dans la Société Française, XVIIIe-XXe siècle* (Paris: Odile Jacob, 1999), 302; François Vallat, *Les Bœufs Malades de la Peste. La Peste Bovine en France et en Europe, XVIIIe-XIXe siècle* (Rennes: Presses Universitaires de Rennes, 2009), 85.

3 Reynal, *Traité de la Police Sanitaire des Animaux Domestiques*, VI–VII.

4 Jürgen Osterhammel, *The Transformation of the World: A Global History of the Nineteenth Century* (Princeton: Princeton University Press, 2014).

5 Ferenc Hutyra and József Marek, *Die Orientalische Rinderpest. Mit besonderer Berücksichtigung der klinischen und anatomischen Merkmale und der Differentialdiagnose* (Jena: Fischer, 1916); Edward Haigh, *Remarks on the Subject of Rinderpest, Russian Plague, the Present Prevailing Disease amongst Cattle* (Manchester: no publisher, 1865). See also: Joanna Swabe, *Animals, Disease, and Human Society. Human-Animal Relations and the Rise of Veterinary Medicine* (London and New York: Routledge, 1999), 73; Thomas Barrett et al., 'Rinderpest – An Old and Worldwide Story: History to c. 1902', in *Rinderpest and Peste des Petits Ruminants. Virus Plagues of Large and Small Ruminants*, ed. Thomas Barrett, Paul-Pierre Pastoret, and William P. Taylor (London and Burlington, MA: Elsevier/Academic, 2006), 87.

6 About European constructions of the 'East' in contexts of epidemics: Patrick Zylberman, 'Civilizing the State. Borders, Weak States and International Health in Modern Europe', in *Medicine at the Border. Disease, Globalization and Security, 1850 to the Present*, ed. Alison Bashford (Houndmills, Basingstoke: Palgrave Macmillan, 2006), 25–9; Peter Baldwin, *Contagion and the State in*

Europe, 1830–1930 (Cambridge, New York: Cambridge University Press, 1999), chapter 'Europe against the Orient', 226–36.

7 Reynal included Austria-Hungary, Russia and Algeria as 'oriental' livestock-exporting states: Reynal, *Traité de la Police Sanitaire des Animaux Domestiques*, VIII.

8 About the use of war-related metaphors in bacteriology and the medical sciences see also Philipp Sarasin, 'Die Visualisierung des Feindes. Über metaphorische Technologien der frühen Bakteriologie', *Geschichte und Gesellschaft* 30, no. 2 (2004): 250–76. On the example of Germany: Silvia Berger, *Bakterien in Krieg und Frieden. Eine Geschichte der medizinischen Bakteriologie in Deutschland 1890–1933* (Göttingen: Wallstein, 2009).

9 Amanda Kay McVety, *The Rinderpest Campaigns: A Virus, Its Vaccines, and Global Development in the Twentieth Century* (Cambridge: Cambridge University Press, 2018), 20. About the 'Hungarian' cattle trade and its consequences in detail: Madeleine Ferrières, *Sacred Cow, Mad Cow: A History of Food Fears* (New York: Columbia University Press, 2006), 177–81.

10 Ferrières, *Sacred Cow, Mad Cow*, 172.

11 Ferrières, *Sacred Cow, Mad Cow*, 175.

12 About the perception of connections between epizootics and epidemics and the fear of zoonotic diseases: Ferrières, *Sacred Cow, Mad Cow*, 193–8, about the coincidences of human and animal diseases: 197.

13 See in detail: Andrew D. Cliff and Matthew Smallman-Raynor, *Oxford Textbook of Infectious Disease Control. A Geographical Analysis from Medieval Quarantine to Global Eradication* (Oxford: Oxford University Press, 2013), 2–18.

14 On the development of quarantine practices: Cliff and Smallman-Raynor, *Infectious Disease Control*, 8–10, 65–75.

15 About epizootics in 18th century Europe and their consequences for the development of veterinary medicine see: Lise Wilkinson, 'Rinderpest and Mainstream Infectious Disease Concepts in the Eighteenth Century', *Medical History* 28, no. 2 (1984): 129–50; Swabe, *Animals, Disease, and Human Society*, 64–71; Ferrières, *Sacred Cow, Mad Cow*, 171–92; Clive A. Spinage, *Cattle Plague: A History* (Dordrecht and London: Kluwer, 2003), 103–20.

16 Spinage, *Cattle Plague*, 105–6; Ferrières, *Sacred Cow, Mad Cow*, 178–80.

17 Ferrières, *Sacred Cow, Mad Cow*, 172–3.

18 Ferrières, *Sacred Cow, Mad Cow*, 185–8.

19 Ferrières, *Sacred Cow, Mad Cow*, 171.

20 Giovanni Maria Lancisi (1654–1720) worked as physician and epidemiologist, especially researching malaria, influenza and rinderpest. About Lancisi's work on animal diseases and the consequences: Wilkinson, 'Rinderpest and Mainstream Infectious Disease Concepts in the Eighteenth Century'; Adriano Mantovani and R. Zanetti, 'Giovanni Maria Lancisi: De Bovilla Peste and Stamping Out', *Historia Medicinae Veterinariae* 18, no. 4 (1993): 97–110; Ferrières, *Sacred Cow, Mad Cow*, 181–8.

21 Barrett et al., 'Rinderpest – An Old and Worldwide Story: History to c. 1902', 89.

22 About the consequences for Europe see also: Swabe, *Animals, Disease, and Human Society*, 63; Mark Harrison, *Contagion: How Commerce Has Spread Disease* (New Haven: Yale University Press, 2012), 213; Ferrières, *Sacred Cow, Mad Cow*, 184–94.

23 Ferrières, *Sacred Cow, Mad Cow*, 185.

24 About the sale of meat from diseased animals in different historical contexts: Ferrières, *Sacred Cow, Mad Cow*, 203–8, 223–30.

25 Ferrières, *Sacred Cow, Mad Cow*, 223.

26 About the foundation of veterinary schools: Lise Wilkinson, *Animals and Disease. An Introduction to the History of Comparative Medicine* (Cambridge and New York: Cambridge University Press, 1992), 65–86; Swabe, *Animals, Disease, and Human Society*, 67–71.

27 Swabe, *Animals, Disease, and Human Society*, 68.
28 Michael Bresalier, Angela Cassidy, and Abigail Woods, 'One Health in History', in *One Health. The Theory and Practice of Integrated Health Approaches*, ed. Jakob Zinsstag et al. (Wallingford, Boston: Cabi, 2015), 4.
29 About the establishment of veterinary schools in several countries: Wilkinson, *Animals and Disease*, 65–85. See also: Daniela Haarmann and Kerstin Weich, 'Geschichte der Tiermedizin', in *Tiere. Kulturwissenschaftliches Handbuch*, ed. Roland Borgards (Stuttgart: Metzler, 2016), 153.
30 The French veterinarian Claude Bourgelat (1712–1779) was, as the founder of the two veterinary schools in Lyon and Alfort, an important figure in the institutionalisation of veterinary medicine. See for Bourgelat and the work of the veterinary schools in France: Wilkinson, *Animals and Disease*, 65–71. On the term epizootic: Bernard Toma, 'A Propos de l'Origine des Mots "épizootique" et "épizootie"', *Bulletin de la Société Française d'Histoire de la Médecine et des Sciences Vétérinaires* 7 (2007): 26–49.
31 In detail about this process: Swabe, *Animals, Disease, and Human Society*, 81–4. Critically about the notion of 'bacteriological revolution': Michael Worboys, 'Was there a Bacteriological Revolution in Late Nineteenth-Century Medicine?', *Studies in History and Philosophy of Biological and Biomedical Sciences* 38, no. 1 (2007): 20–42.
32 For the work of Robert Koch in the scientific context of his time: Christoph Gradmann, *Krankheit im Labor: Robert Koch und die medizinische Bakteriologie* (Göttingen: Wallstein, 2005).
33 About Pasteur's developments of animal vaccinations: M. Lombard, P. P. Pastoret, and A. M. Moulin, 'A Brief History of Vaccines and Vaccination', *Revue Scientifique et Technique (OIE)* 26, no. 1 (2007): 32–5. See also: Swabe, *Animals, Disease, and Human Society*, 81–3.
34 Alan L Olmstead, 'The First Line of Defense: Inventing the Infrastructure to Combat Animal Diseases', *The Journal of Economic History* 69, no. 2 (2009): 332.
35 Fred Brown, 'The History of Research in Foot-and-Mouth Disease', *Virus Research* 91, no. 1 (2003): 3–4.
36 Türel Özkul and R. Tamay Başagaç Gül, 'La Collaboration de Maurice Nicolle et Adil Mustafa. La Découverte de l'Agent de la Peste Bovine', *Revue de Médecine Vétérinaire* 159, no. 4 (2008): 243–6.
37 Swabe, *Animals, Disease, and Human Society*, 81.
38 Axel Hüntelmann, 'Geschichte des Tierversuchs', in *Tiere. Kulturwissenschaftliches Handbuch*, ed. Roland Borgards (Stuttgart: Metzler, 2016), 166–7.
39 Hüntelmann, 'Geschichte des Tierversuchs', 169.
40 In detail about animal experiments in the late 19th and early 20th centuries and the opponents: Anita Guerrini, *Experimenting with Humans and Animals. From Galen to Animal Rights* (Baltimore: Johns Hopkins University Press, 2003), 93–113.
41 On European concepts: Adriano Mantovani, 'Notes on the Development of the Concept of Zoonoses', *Historia Medicinae Veterinaria* 26, no. 2 (2001): 41–2.
42 About popular fears of zoonotic diseases and scientific discourses: Ferrières, *Sacred Cow, Mad Cow*, 193–216.
43 Mantovani, 'Notes on the Development of the Concept of Zoonoses', 42.
44 About Pierre François Olive Rayer (1793–1867) and his work on glanders: Lise Wilkinson, 'Glanders: Medicine and Veterinary Medicine in Common Pursuit of a Contagious Disease', *Medical History* 25, no. 4 (1981): 379-84.
45 Swabe, *Animals, Disease, and Human Society*, 84–5. See also for an overview: James H. Steele, 'A Bookshelf on Veterinary Public Health', *American Journal of Public Health* 63, no. 4 (1973): 291–311. Examples of early and mid-19th century research: Bernard Anton Greve, *Erfahrungen und Beobachtungen*

über die Krankheiten der Hausthiere im Vergleich mit den Krankheiten der Menschen, 2 vols (Oldenburg: Schulze, 1818); Jean-Baptiste-Hippolyte Veyssière, *Des Maladies Transmissibles des Animaux à l'Homme* (Paris: Libraires de l'École de Médecine, 1852). See also: Mantovani, 'Notes on the Development of the Concept of Zoonoses', 44. About debates on the borderlines between human beings and animals: Anne Hardy, 'Animals, Disease, and Man: Making Connections', *Perspectives in Biology and Medicine* 46, no. 2 (2003): 202–3.

46 Rudolf Virchow, 'Intoxicationen, Zoonosen und Syphilis', in *Handbuch der speciellen Pathologie und Therapie*, ed. Frieder Alexander Simon, Carl P. Falck, and Rudolf Virchow, vol. 2, 1 (Erlangen: F. Enke, 1855), 337–8 (translation of citation: CK). See also: Mantovani, 'Notes on the Development of the Concept of Zoonoses', 44. Briefly about the etymological background: Hardy, 'Animals, Disease, and Man: Making Connections', 202; Richard N. Fiennes, *Zoonoses and the Origins and Ecology of Human Disease* (London and New York: Academic Press, 1978), X–XII.

47 Wilhelm Probstmayr, *Etymologisches Wörterbuch der Veterinär-Medicin und ihrer Hilfswissenschaften* (München: Grubert, 1863), 844.

48 Bruno Galli-Valerio, *Zoonosi: Malattie Trasmissibili dagli Animali all'Uomo* (Milano: Ulrico Hoepli, 1894). See also: Mantovani, 'Notes on the Development of the Concept of Zoonoses', 44.

49 World Health Organization, *Joint WHO/FAO Expert Committee on Zoonoses. Second Report, Stockholm, 11–16 August 1958*, World Health Organization Technical Report Series 169 (Geneva: World Health Organization, 1959), 6.

50 R. Bauerfeind et al., *Zoonosen: Von Tier zu Mensch übertragbare Infektionskrankheiten* (Köln: Deutscher Ärzteverlag, 2013), 1.

51 Lombard, Pastoret, and Moulin, 'A Brief History of Vaccines and Vaccination', 29.

52 David Jones and Stefan Helmreich, 'A History of Herd Immunity', *The Lancet* 396, no. 10254 (September 2020): 810.

53 Lombard, Pastoret, and Moulin, 'A Brief History of Vaccines and Vaccination', 30.

54 Lombard, Pastoret, and Moulin, 'A Brief History of Vaccines and Vaccination', 30.

55 Hervé Bazin, 'L'Histoire des Vaccinations. 1ère Partie: De la Variolisation à la Vaccination', *Bulletin de la Société Française d'Histoire de la Médecine et des Sciences Vétérinaires* 12 (2012): 143.

56 Angela von den Driesch and Peter Joris, *Geschichte der Tiermedizin. 5000 Jahre Tierheilkunde* (Stuttgart: Callwey, 2002), 180–1; Lombard, Pastoret, and Moulin, 'A Brief History of Vaccines and Vaccination', 38; Bazin, 'De la Variolisation à la Vaccination', 141–2.

57 About vaccination against smallpox as a public health strategy: Baldwin, *Contagion and the State in Europe, 1830–1930*, 244–54, especially 245–6; Guerrini, *Experimenting with Humans and Animals*, 67–9.

58 Lombard, Pastoret, and Moulin, 'A Brief History of Vaccines and Vaccination', 38.

59 Lombard, Pastoret, and Moulin, 'A Brief History of Vaccines and Vaccination', 30.

60 For an overview: Swabe, *Animals, Disease, and Human Society*, 83.

61 On the history of anthrax: Susan D. Jones, *Death in a Small Package. A Short History of Anthrax* (Baltimore: Johns Hopkins University Press, 2010); Driesch and Joris, *Geschichte der Tiermedizin*, 182–4. See also Chapter 5.

62 On anthrax vaccine development: Lombard, Pastoret, and Moulin, 'A Brief History of Vaccines and Vaccination', 34; Jones, *Death in a Small Package*, 66–76.

63 Kathleen Kete, 'La Rage and the Bourgeoisie: The Cultural Context of Rabies in the French Nineteenth Century', *Representations* 22, no. 1 (1988): 90–2.

64 Bert Hansen, 'America's First Medical Breakthrough: How Popular Excitement about a French Rabies Cure in 1885 Raised New Expectations for Medical Progress', *American Historical Review* 103 (1998): 373–418.

65 Lombard, Pastoret, and Moulin, 'A Brief History of Vaccines and Vaccination', 35. See also Chapter 5.

66 Lombard, Pastoret, and Moulin, 'A Brief History of Vaccines and Vaccination', 43; Scott McVey and Jishu Shi, 'Vaccines in Veterinary Medicine: A Brief Review of History and Technology', *Veterinary Clinics of North America: Small Animal Practice* 40, no. 3 (2010): 382. About vaccine research at the BAI: Olmstead, 'The First Line of Defense'.

67 Lombard, Pastoret, and Moulin, 'A Brief History of Vaccines and Vaccination', 37–8.

68 Lombard, Pastoret, and Moulin, 'A Brief History of Vaccines and Vaccination', 37–8; McVey and Shi, 'Vaccines in Veterinary Medicine', 382. For biographical information on Gaston Ramon (1886–1963): Institut Pasteur, 'Institut Pasteur: Gaston Ramon – Pro Humanitate', accessed April 1, 2021, https://www.pasteur.fr/fr/institut-pasteur/notre-histoire/gaston-ramon-pro-humanitate.

69 Lombard, Pastoret, and Moulin, 'A Brief History of Vaccines and Vaccination', 39–43.

70 For an overview on early 20th century efforts of rinderpest vaccination: William P. Taylor, Peter L. Roeder, and Mark M. Rweyemamu, 'History of Vaccines and Vaccination', in *Rinderpest and Peste des Petits Ruminants. Virus plagues of large and small ruminants*, ed. William P. Taylor, Thomas Barrett, and Paul-Pierre Pastoret (London and Burlington, MA: Elsevier/Academic, 2006), 223–5.

71 Lombard, Pastoret, and Moulin, 'A Brief History of Vaccines and Vaccination', 39.

72 McVey and Shi, 'Vaccines in Veterinary Medicine', 382.

73 Swabe, *Animals, Disease, and Human Society*, 83–4, 86–7.

74 Lombard, Pastoret, and Moulin, 'A Brief History of Vaccines and Vaccination', 33. About veterinarians and innovations in microbiology see also, with the example of Britain: Michael Worboys, 'Germ Theories of Disease and British Veterinary Medicine, 1860–1890', *Medical History* 35 (1991): 308–27.

75 Anne Hardy, 'Pioneers in the Victorian Provinces: Veterinarians, Public Health and the Urban Animal Economy', *Urban History* 29, no. 3 (2002): 372–87; Worboys, 'Germ Theories of Disease and British Veterinary Medicine, 1860–1890', 318–19.

76 About the context: Justin Kastner, 'Food and Agriculture Security: An Historical Illustration of Contemporary Challenges', in *Food and Agriculture Security: An Historical, Multidisciplinary Approach* (Santa Barbara: Praeger, 2011), 5–7; Swabe, *Animals, Disease, and Human Society*, 71–2; Harrison, *Contagion*, 211–12.

77 For the link between the growth of population, here in case of Europe, and the increasing food importations in the 19th century: Swabe, *Animals, Disease, and Human Society*, 71–2.

78 James Troubridge Critchell and Joseph Raymond, *A History of the Frozen Meat Trade* (London: Constable, 1912), 16.

79 Lukas Nieradzik, 'Geschichte der Nutztiere', in *Tiere. Kulturwissenschaftliches Handbuch*, ed. Roland Borgards (Stuttgart: Metzler, 2016), 126.

80 Harrison, *Contagion*, 214; Swabe, *Animals, Disease, and Human Society*, 72–3.

81 About the British rinderpest epizootic of 1866: Harrison, *Contagion*, 216–19; Swabe, *Animals, Disease, and Human Society*, 74–8. See also: John R. Fisher, 'Cattle Plagues Past and Present: The Mystery of Mad Cow Disease', *Journal of Contemporary History* 33, no. 2 (1998): 215; Spinage, *Cattle Plague*, 275–9.

82 On occasion of the epizootic of rinderpest in France, the French Ministry of Agriculture and Commerce edited a collection of all existing French legislation concerning rinderpest since 1714: République Française, Ministère de l'Agriculture et du Commerce, ed., *Documents sur la Peste Bovine* (Paris: Imprimerie Nationale, 1872). See also: Bertrand Taithe, *Defeated Flesh: Medicine, Welfare, and Warfare in the Making of Modern France* (Manchester: Manchester University Press, 1999), 121.

83 Barrett et al., 'Rinderpest – An Old and Worldwide Story: History to c. 1902', 99–100.

84 About epizootics in colonial India: Saurabh Mishra, 'Beasts, Murrains, and the British Raj: Reassessing Colonial Medicine in India from the Veterinary Perspective, 1860–1900', *Bulletin of the History of Medicine*, 2011, 587–619; Samiparna Samanta, 'Dealing with Disease: Epizootics, Veterinarians and Public Health in Colonial Bengal, 1850–1920', in *Medicine and Colonialism: Historical Perspectives in India and South Africa*, ed. Poonam Bala (Routledge, 2015), 61–74. Detailed about rinderpest in India in the 19th century: Spinage, *Cattle Plague*, 447–84.

85 About Texas fever of cattle in the US: Alan L. Olmstead and Paul W. Rhode, *Arresting Contagion: Science, Policy, and Conflicts over Animal Disease Control* (Cambridge, Mass.: Harvard University Press, 2015), Chapter 11; Harrison, *Contagion*, 222–4.

86 McVety, *The Rinderpest Campaigns*, 24–9. The epizootics of rinderpest and other animal diseases in Africa have during the last decades come much into the focus of historical research, revealing a multiplicity of impacts on colonial and postcolonial societies. See for instance: Pule Phoofolo, 'Face to Face with Famine: The BaSotho and the Rinderpest, 1897–1899', *Journal of Southern African Studies* 29, no. 2 (June 1, 2003): 503–27; Daniel Gilfoyle, 'Veterinary Research and the African Rinderpest Epizootic: The Cape Colony, 1896–1898', *Journal of Southern African Studies* 29, no. 1 (2003): 133–54; Karen Brown, 'Tropical Medicine and Animal Diseases: Onderstepoort and the Development of Veterinary Science in South Africa, 1908–1950', *Journal of Southern African Studies* 31, no. 3 (2005): 513–29; Thaddeus Sunseri, 'The Entangled History of Sadoka (Rinderpest) and Veterinary Science in Tanzania and the Wider World, 1891–1901', *Bulletin of the History of Medicine* 89, no. 1 (2015): 95–121.

87 For an overview and discussion of historiography see: Thaddeus Sunseri, 'The African Rinderpest Panzootic, 1888–1897', Oxford Research Encyclopedia of African History, April 26, 2018, https://oxfordre.com/africanhistory/view/10.1093/acrefore/9780190277734.001.0001/acrefore-9780190277734-e-375.

88 George Fleming, *A Manual of Veterinary Sanitary Science and Police*, vol. 1 (London: Chapman & Hall, 1875), XVIII. The Scottish military veterinarian George Fleming (1833–1901) published numerous books on diverse topics of animal health; about Fleming and his *Manual* see also: Hardy, 'Pioneers in the Victorian Provinces', 379–80.

89 Hardy, 'Pioneers in the Victorian Provinces', 379–80.

90 Fleming, *A Manual of Veterinary Sanitary Science*, 1: 216–17.

91 Nieradzik, 'Geschichte der Nutztiere', 125–6.

92 Nieradzik, 'Geschichte der Nutztiere', 126. For a detailed contemporary description: Article 'Slaughterhouses', in: Alexander Wynter Blyth, ed., *A Dictionary of Hygiène and Public Health, Comprising Sanitary Chemistry, Engineering, and Legislation, the Dietetic Value of Foods, and the Detection of Adulterations* (London: C. Griffin and Company, 1876), 541–7.

93 For the history of slaughterhouses: Amy J. Fitzgerald, 'A Social History of the Slaughterhouse: From Inception to Contemporary Implications', *Human Ecology Review* 17, no. 1 (2010): 58–69; Dorothee Brantz, 'Slaughter in the City. The Establishment of Public Abattoirs in Paris and Berlin, 1780–1914' (PhD Thesis, Chicago, University of Chicago, 2003); Christopher Otter, 'Civilizing Slaughter: The Development of the British Public Abattoir, 1850–1910', *Food and History* 3, no. 2 (2008): 29–51; Paula Young Lee, ed., *Meat, Modernity, and the Rise of the Slaughterhouse* (Durham, NH: University of New Hampshire Press, 2008).

94 Upton Sinclair, *The Jungle* (New York: Doubleday, Page & Co., 1906); about the context see: Fitzgerald, 'A Social History of the Slaughterhouse', 60; Dominic A. Pacyga, 'Chicago: Slaughterhouse to the World', in: *Meat, Modernity, and the Rise of the Slaughterhouse*, ed. Paula Young Lee (Durham, NH: University of New Hampshire Press, 2008), 153–66.

95 Nieradzik, 'Geschichte der Nutztiere', 126.
96 About the development of the globalised meat trade: Richard Perren, *Taste, Trade and Technology: The Development of the International Meat Industry since 1840* (Aldershot: Ashgate, 2006).
97 John Edwin Wrenn and Frank M. Surface, *International Trade in Meats and Animal Fats* (Washington: U.S. Govt. Printing Office, 1925), 2, 320–1.
98 For the increased significance of veterinary public health in the late 19th century see also: Swabe, *Animals, Disease, and Human Society*, 84–7.
99 About the range of diseases of relevance for meat inspectors during the 1890s and 1900s see: Richard H. Edelmann, 'Fleischbeschau', in *Handbuch der Hygiene*, ed. Theodor Weyl, vol. 3 (Jena: Gustav Fischer, 1896), 475–546. A historiographical overview on the problems of veterinary inspection (with a focus on the system of the Netherlands): Peter A. Koolmees, 'Veterinary Inspection and Food Hygiene in the Twentieth Century', in *Food, Science, Policy, and Regulation in the Twentieth Century: International and Comparative Perspectives*, ed. Jim Phillips and David F. Smith (London and New York: Routledge, 2000), 53–68. On the influences on the veterinary profession see also: Swabe, *Animals, Disease, and Human Society*, 86–7.
100 About the trichinosis outbreaks in Germany and the cultural implications: Dorothee Brantz, 'How Parasites Make History: On Pork and People in Nineteenth-Century Germany and the United States', *Bulletin of the GHI Washington*, no. 36 (2005): 72; Patrick Zylberman, 'Making Food Safety an Issue: Internationalized Food Politics and French Public Health from the 1870s to the Present', *Medical History*. 48, no. 1 (2004): 6–7; Uwe Spiekermann, 'Dangerous Meat? German-American Quarrels over Pork and Beef', *Bulletin of the GHI Washington* 46 (Spring 2010): 95–6.
101 On the process of introducing meat inspection by state authorities see also: Swabe, *Animals, Disease, and Human Society*, 85–7.
102 Robert von Ostertag, *Handbook of Meat Inspection*, 2nd ed. (Baillière & Co.: London, 1905); Calvin W. Schwabe, *Veterinary Medicine and Human Health*, 2nd ed. (Baltimore: Williams & Wilkins, 1984), 46. Robert von Ostertag (1864–1940), veterinary professor in Stuttgart and Berlin, was a vigorous advocate of scientifically-based systems of meat inspection in Germany. See: Eberhard Bulling and Arno Schönberg, 'Robert von Ostertag (1864–1940): A Veterinarian Contemporary with R. Virchow and R. Koch', *Historia Medicinae Veterinariae* 24, no. 4 (1999): 97–119.
103 Robert von Ostertag, *Bibliographie der Fleischbeschau. Zugleich Ergänzung zum Handbuch der Fleischbeschau* (Stuttgart: F. Enke, 1905).
104 Zylberman, 'Making Food Safety an Issue', 6–11.
105 For research on the so-called pork war see: Justin Kastner, 'Managing Human and Animal-Health Threats: Additional Lessons from the 19th-Century Trading World', in *Food and Agriculture Security: An Historical, Multidisciplinary Approach* (ABC-CLIO, 2011), 106–11; Olmstead, 'The First Line of Defense', 342–9; Brantz, 'How Parasites Make History'; Spiekermann, 'Dangerous Meat?'; Zylberman, 'Making Food Safety an Issue', 6–11; Suellen Hoy and Walter Nugent, 'Public Health or Protectionism? The German-American Pork War, 1880–1891', *Bulletin of the History of Medicine* 63 (1989): 198–224.
106 Harrison, *Contagion*, 226.
107 *Deutsche Vierteljahrsschrift für öffentliche Gesundheitspflege. Organ des Deutschen Vereins für öffentliche Gesundheitspflege*, 1 (1869)–47 (1915)/*Jahresbericht über die Fortschritte und Leistungen auf dem Gebiet der Hygiene*, 1 (1883/1884)–31 (1913/1915).
108 About anthrax as an occupational disease see: Jones, *Death in a Small Package*, 81–127. See also Chapter 5.

109 For an overview on the history of bovine tuberculosis: Walter Bisping, 'Zur Geschichte der Bekämpfung der Rindertuberkulose, *Berliner und Münchener Tierärztliche Wochenschrift* 111, no. 10 (1998): 362–7. A detailed study about bovine tuberculosis in the British context: Keir Waddington, *The Bovine Scourge: Meat, Tuberculosis and Public Health, 1850–1914* (Woodbridge: Boydell Press, 2006). For the U.S. Alan L. Olmstead and Paul W. Rhode, 'The "Tuberculous Cattle Trust": Disease Contagion in an Era of Regulatory Uncertainty', *The Journal of Economic History* 64, no. 4 (2004): 929–63; Alan L. Olmstead and Paul W. Rhode, 'An Impossible Undertaking: The Eradication of Bovine Tuberculosis in the United States', *Journal of Economic History* 64, no. 3 (2004): 734–72. On Europe: Barbara Orland, 'Handeln in Zeiten der Ungewißheit. Tuberkulose, Milch und Tierseuchenbekämpfung im 19. und 20. Jahrhundert', *Mittelungen des Internationalen Arbeitskreises für Kulturforschung des Essens* 8 (May 2001): 13–24.

110 In detail: Orland, 'Handeln in Zeiten der Ungewißheit', 15–8.

111 Driesch and Joris, *Geschichte der Tiermedizin*, 177–8; Orland, 'Handeln in Zeiten der Ungewißheit', 14–17; Bisping, 'Bekämpfung der Rundertuberkulose', 362.

112 Edelmann, 'Fleischbeschau', 497; see also: Orland, 'Handeln in Zeiten der Ungewißheit', 18; Haarmann and Weich, 'Geschichte der Tiermedizin', 155.

113 In detail about the discussions in its multiple facets: Orland, 'Handeln in Zeiten der Ungewißheit', 17–22.

114 Milton J. Rosenau, 'The Number of Bacteria in Milk and the Value of Bacterial Counts', in *Milk and Its Relation to the Public Health*, ed. U.S. National Institutes of Health (Washington: Government Printing Office, 1908), 421–55. See also: Ferrières, *Sacred Cow, Mad Cow*, 289–92.

115 About the history of brucellosis see: Jean Théodoridès, 'Bref Historique de la Fièvre de Malte', *Histoire des Sciences Médicales* 30, no. 1 (1996): 87–90. John Rizzo Naudi, *Brucellosis – The Malta Experience: A Celebration, 1905–2005* (San Gwann (Malta): Publishers Enterprise Group (PEG), 2005); E. M. O'Hern, 'Alice Evans and the Brucellosis Story', *Annali Sclavo: Rivista di Microbiologia e di Immunologia* 19, no. 1 (1977): 12–9.

116 Orland, 'Handeln in Zeiten der Ungewißheit', 23–4.

117 Orland, 'Handeln in Zeiten der Ungewißheit', 24.

118 See also: Swabe, *Animals, Disease, and Human Society*, 86–7.

3 International approaches to contain cross-border epizootics, 1863–1914

In July 1863, the first International Veterinary Congress convened in Hamburg. The timing was significant—it was a decade of intensifying global connections through innovations in communication technologies and traffic. The expansion of the telegraph network made the transfer of information across continents and oceans radically faster. In 1866, the first transatlantic telegraph cable was successfully installed.[1] New steamship and railway lines were established for travel and transport across large distances. The opening of the Suez Canal in 1869 provided a considerably faster traffic connection between Europe, Asia and Africa for passengers and the shipping of commodities. Related to these developments, the dynamics of the 1860s also fueled new developments in internationalism.[2] For many contemporaries, there was a clear link between the increasingly globalised trade and traffic connections and the spread of animal diseases across borders and oceans. The trade with animals and animal products was hampered or sometimes completely interrupted by veterinary control measures, and for some governments, the discussions on the animal disease situation in exporting countries served as an argument to completely block the trade with animals and animal products. Not surprisingly, the issue of animal diseases was one of the first areas to be addressed by projects of internationalism since the early 1860s by governments and veterinary experts.

At the first meetings of veterinary internationalism, experts began to discuss concepts and practices that centred on the problem of the cross-border spread of animal diseases. They explored opportunities to set up a new international agenda of disease control by interacting with diplomats and governmental representatives, while strengthening their scientific community and professional authority through meeting and networking at international platforms. For developing international strategies, they oriented themselves on their respective national sanitary systems as well as on international strategies in other areas, for example in the fields of public health and plant diseases. Over several decades, they discussed and further elaborated their ideas and concepts, but there was a significant impediment in these efforts: when the Tenth International Veterinary Congress began in London in August 1914, coinciding with the outbreak of the First World War, the international meetings had neither resulted in a centralised

DOI: 10.4324/9781003033998-3

international system of animal disease information nor in an official multilateral agreement concerning long-term and permanent cooperation efforts. At least there existed a closely networked scientific community of experts who had discussed and established in detail an instrumentarium for future international anti-disease strategies. These concepts and practices to control the spread of animal diseases, even if not used in practice at the international level during these decades, would become the basis of future activities in the interwar years and later. Experts active on international platforms managed to create a long-term awareness of the global and cross-border dimension of animal disease problems, not only in their own scientific community but gradually also among governmental policy makers. This process of defining and developing an international instrumentarium to address animal disease problems was ultimately linked to questions of power: the instruments of animal disease control discussed on international platforms were primarily aimed to protect European countries from diseases and to advance the 'Europeanisation' of the world, in this case in the field of veterinary medicine.

Hamburg, Vienna, Zurich: international veterinary congresses in the 1860s

Several factors contributed to the development of veterinary internationalism in the second half of the 19th century. As Madeleine Herren has outlined, the 1860s were a decisive period for internationalism, with international meetings and organisations starting to take shape in new and stabilised structural forms, increasing in numbers, and new actors taking up activities on the international stage.[3] Organising international meetings became logistically far easier than before. Events such as the new world exhibitions that took place since the 1850s not only initiated important infrastructure projects but also attracted the organisers of international congresses.[4] Scientific professionals from various fields and branches founded international associations to foster regular scientific exchange.[5] Various governments showed interest in more active participation in technical and scientific international projects. On the basis of multilateral conventions, governments cooperated to establish a series of international commissions and so-called public international unions, the forerunners of international governmental organisations, such as the International Telegraph Union in 1865.[6] They served as regulatory bodies for areas of common international concern or to strengthen European interests. Among the early fields of international engagement was the area of epidemics control, linked with the aim of European governments to prevent the spread of epidemics to European countries. Following diplomatic pressure from several European governments, sanitary councils outside Europe were founded in Alexandria (1831), Constantinople (1838), Tangier (1840) and Teheran (1881) with international delegations to monitor the surveillance of sanitary measures.[7] In 1851, on the initiative of the French government, the first of the series of International Sanitary Conferences discussed joint measures against cholera, especially quarantine regulations.[8] The participants of the conferences included diplomats and medical experts; they aimed to consolidate, as the French epidemiologist Adrien Proust phrased it in the 1890s, "the defence of Europe"[9] against contagious diseases such as cholera and plague.

In the field of epizootics control, it was a veterinarian who took the first step to initiate international meetings.[10] John Gamgee, veterinary professor and founder of the New Edinburgh Veterinary College, was frustrated by what he considered an inadequate animal disease control system in Britain and advocated for stricter procedures.[11] He repeatedly warned about the importation of contagious animal diseases and had acquired a reputation for challenging conventional scientific opinions.[12] Gamgee argued that the natural protection of the British Isles from disease importation had been relinquished in the interests of free trade—opening the markets to free trade in animals would lead to flows of epizootics.[13] In the spring of 1863, he began to advance his cause by contacting colleagues on the European continent. In cooperation with Eduard Hering, professor at the veterinary school in Stuttgart, Gamgee sent a circular letter inviting professional veterinarians to an international meeting and published the invitation in veterinary journals.[14] Gamgee's circular summarised the challenges presented by the spread of epizootics in mid-19th–century Europe in relation to the developments in trade and transport. He argued that European governments had a twofold problem with regard to the spread of contagious animal diseases—the importation of potentially infected animals in huge numbers without adequate control systems, and the confusing lack of accurate information about the disease situation in many trading countries. Gamgee appealed to the economic interests of European national governments.[15] He asked his colleagues to provide information on the animal disease situation in their countries or districts, on the issue of import and export of livestock and the direction of spread of the related animal diseases, as well as on their experience with effective disease control measures. The objective of the new international meeting of veterinarians was to suggest to the governments of all European countries the establishment of a common set of measures to prevent the spread of animal diseases. The circular pointed out that veterinarians were in an important position of expertise with their specialised knowledge in this important field of commercial relations, and that they had an obligation to support economic progress by containing epizootics and preventing food shortages.[16]

The result of Gamgee's initiative was the first International Veterinary Congress taking place in July 1863 as a meeting of experts. To attract as many colleagues as possible, the organisers chose Hamburg as the congress venue. The city not only had excellent traffic connections, but was also in the same month hosting the International Agricultural Exhibition, which presented the latest trends in agricultural technology and livestock breeding.[17] Gamgee's idea was successful: about 100 veterinarians were listed as participants in the first international congress.[18] This was a promising start and reflected the broad interest in the area. The experts came from German countries, Austria, Switzerland, Luxemburg, Norway, Sweden, Denmark, Italy, Russia and, with Gamgee and one additional colleague, from Britain. Already by their large number, the German-speaking experts dominated the congress. The list of participants included many renowned veterinarians and scientists, among them the professors Moritz Friedrich Röll from Vienna, Carl Heinrich Hertwig from Berlin, Gottlieb Carl Haubner from Dresden, Hans Rudolf Zangger from Zurich and Christian Gerlach from Hanover.[19] The range of participants also showed the difficulties of international meetings in the middle of the 19th century—as the congress report stated, two veterinarians from Hungary and Ukraine arrived when the meeting was already over.[20] The congress focused especially on the issues of rinderpest and bovine pleuropneumonia because of their topicality, significance for the livestock sector, and possible rapid spread across borders.[21] The starting point was a discussion on the shortcomings of the veterinary sanitary border systems and the associated quarantine regulations with regard to the livestock exportations of several countries, especially Russia.[22] Experts argued about the duration and strictness of the quarantine regulations. The Viennese veterinary professor Röll explained in an introductory report on rinderpest the dilemma of permeable borderlines and the insufficiency of quarantine practices—the longer and more rigorous the quarantine regulations, the greater the incentive to smuggle animals illegally, he argued, because quarantine times reduced the economic value of animals.[23] There was uncertainty about the details of how the transmission of the disease happened, and through which substances, which complicated the discussion on border control measures and sanitary practices. Another main problem was the uncertainty about the incubation and contagion periods, discussed in the case of rinderpest, as it was unclear whether they corresponded with quarantine durations.[24] Some participants at the congress argued in favour of abolishing strict sanitary measures at the borders, considering these no longer up to date to prevent diseases from spreading.[25] In the discussions about the outbreaks of bovine pleuropneumonia, the participants likewise reported an array of uncertainties to cope with—the incubation and contagion period, the question of culling and its dimensions, the sale of meat of diseased animals, the effectiveness of inoculation experiments, and the economic as well as public health consequences of epizootics, all of which were difficult to be assessed.[26]

The participants of the meeting brought another central issue to the fore: the considerable differences between the various national veterinary control systems, which made sanitary border practices indispensable in the first place. Christian Joseph Fuchs, veterinary professor from Heidelberg University and part of a commission which reformed the veterinary system of the Grand Duchy of Baden, underlined that the international meeting presented a chance to work in general towards uniform principles of veterinary services. Many state administrations, he argued, were in the process of establishing or reforming their veterinary sanitary systems; if they agreed from the start to introduce similar structures of veterinary control, some of the uncertainties about disease situations could be reduced.[27] This idea of creating uniform national veterinary systems, as far as possible, was also in future discussions until the late 19th century often included as an element of a new and trustworthy system of animal disease control. However, what actually defined a veterinary system as effective and trustworthy would take up much of the discussions at further international congresses. In order to move their cause forward, the congress participants drafted a series of resolutions which were submitted as recommendations to their respective governments. They agreed on a list of diseases for future international attention, including several which could be transmitted to humans—rabies, anthrax, glanders and skin worms, foot-and-mouth disease, scabies, rinderpest, bovine pleuropneumonia, sheep pox, sheep foot-and-mouth disease, dourine of horses.[28] Further, they underlined the need to introduce state-organised veterinary services in all countries as a prerequisite for any cooperation, together with a centralised disease statistics register and a regular and common monitoring system for rail animal transport.[29] Hering, the meeting's president, concluded that it was now the task of the governments to proceed towards concrete arrangements and implementations.[30] The result of the Hamburg meeting was twofold. First, a number of essential points regarding the spread of animal diseases in its cross-border dimension were defined and communicated to the attention of governments. Second, the event fostered the creation of a new professional community of—at this time mainly European—veterinary experts, who continued activities of scientific exchange and the discussion on international measures concerning cross-border animal diseases.

Already in 1865, a Second International Veterinary Congress convened in Vienna, focusing again on the topic of veterinary sanitary questions of animal disease control. By that time, the epizootic situation in Europe had clearly deteriorated.[31] The congress coincided with the first months of a severe rinderpest epizootic in Britain and parts of the European continent. The consequences for the British cattle economy were disastrous.[32] Rinderpest also appeared in several other European countries. Austria-Hungary had to deal repeatedly with rinderpest outbreaks introduced by cattle importations from Russia, with severe losses of animals between 1849 and 1865.[33] The situation of epizootics in Europe consolidated the international meeting. The Austrian government supported the congress by formally inviting its European counterparts to send delegations of experts; a financial contribution from the Austrian Emperor covered most of the expenses.[34] The number of participants had more than doubled, and the list included now delegates from Austria-Hungary, the German states, Russia, Great-Britain, the Danube princely states, Denmark, France, Belgium, Italy, Norway, Austria, Portugal, Sweden, Switzerland, Turkey (as part of the Ottoman Empire) and (Russian) Poland.[35] In continuation of the Hamburg event, the aim of the second congress was formulated as "to deliberate on subjects of veterinary science, which are of international importance, to propose motions with reference to them, and to submit resolutions which have been adopted to the separate Governments".[36] In its programme, the congress took up and extended the themes of the Hamburg meeting. It continued the previous debate on rinderpest, especially the problem of its incubation period in relation to border controls and quarantine regulations, and the general discussion about the ways of handling animal trade and traffic problems. The quarantine period for livestock to prevent the introduction of rinderpest could be reduced to 10 days, the resolutions stated, subject to strict and detailed veterinary control measures.[37] The congress also addressed the question of the effectiveness and the risks of vaccination experiments. Furthermore, it included a more detailed discussion on rabies, considered as an increasing problem in many parts of the world. The experts agreed on international principles to contain the disease through the control of dogs. The project suggested listing dogs in central municipal registers for identification, introducing taxes on dog ownership and reducing the number of stray dogs. In as many countries as possible, sanitary services should establish mandatory reporting of cases of rabid animals.[38] This proposal was too ambitious to be realised, but the discussion showed the gradual extension to include diseases relevant to human health, and the idea of permanently linking the local, national and international levels of disease control.

The discussions at the subsequent Third International Veterinary Congress in Zurich in 1867 revealed again how the increasing trade in livestock and animal products confronted European governments with the need for an international agreement. The event, organised with the support of the Swiss government, attracted 188 participants from Austria, the German states, Switzerland, Russia, Great Britain, Italy, France, Belgium, Norway, the Netherlands, Romania, Serbia and Turkey.[39] The outbreaks of rinderpest in Great Britain, the Netherlands, France, Germany and Switzerland were among the topics particularly discussed. The Swiss government representative, the Federal Councilor Carl Schenk, summarised in his opening speech the problematic consequences of connected modern traffic and transport systems:

> It is the peculiarity of our times that, on the one hand, the respective countries are more at risk than before of having their livestock attacked by sudden outbreaks of animal diseases; on the other hand, they are forced to apply protective methods that lead to certain constraints that were not so necessary in earlier times. The great new means of transport of our century makes it possible to transmit diseases from far away and spread them rapidly in new directions. The same powerful means of transport has created an interwoven traffic on multiple levels, and any disturbance for reasons of protection will become itself a precarious vexation; a total interruption of the lines of traffic by blockages will cause extremely serious and noticeable damage to all sides.[40]

This speech highlighted the dilemma of the previous international discussions on the topic—the conventional national measures such as the sanitary border controls were rarely sufficient, severely hampered the flow of trade, and they were therefore a source of frequent conflicts on different levels. The president of the congress, the Zurich veterinary professor Rudolf Zangger, explained that the current outbreaks of epizootics in Europe only confirmed the view of veterinary experts about the necessity for increased governmental action concerning an international coordination of disease control.[41] Among the topics of discussion were again the rinderpest outbreaks, but also the handling of bovine pleuropneumonia, the question of meat inspection, of veterinary education and of the adequate organisation of state veterinary services. A sense of the increasingly urgent need for new forms of international cooperation to improve this situation was in the air. The participants of the congress confirmed in their resolutions the results of the previous congress on rinderpest, but also decided to take a step further in this direction and included a paragraph which requested the Russian government to take a more active role in the future. Since the Hamburg congress, the experts on international platforms indicated the export of Russian animals as a main source of the spread of rinderpest to Central and Western Europe. The paragraph demanded that the Russian government should organise an international veterinary commission tasked to investigate the causes and the geographical seats of the epizootic.[42] The congress resolutions also included assessments on addressing bovine pleuropneumonia as well as a focus on veterinary public health. In particular, the participants demanded a professional control of meat hygiene and recommended introducing a trichinae inspection system with microscopes in slaughterhouses.[43] The participants also voted again to recommend the establishment of regular state-funded veterinary services.[44] By the late 1860s, therefore, the first three international veterinary congresses had produced a number of impulses which would shape the future development of concepts and strategies in veterinary internationalism. The focus of the congresses was clearly on Western and Central European concerns. Several experts had managed to gradually gain the support of some governments for parts of their agenda, especially in the case of Austria and Switzerland. The newly created community of veterinarians legitimised itself by the semi-official position of their members, the power of their professional authority and through the expert resolutions of their international meetings on scientific and veterinary questions, which were communicated to the respective governments in the form of recommendations.

The Vienna rinderpest conference 1872

The International Veterinary Congress planned for the year 1870 in Belgium was cancelled because of the Franco-German War of 1870–1871.[45] The series of congresses continued only in 1883 in Brussels. In the meantime, however, the issue of the cross-border spread of animal diseases, more precisely the case of rinderpest in Europe, became the subject of a governmental international initiative. The Russian government did not take up the international activity suggested in the resolutions of the 1867 congress at Zurich. Instead, in March/April 1872, the Austrian government organised an international conference in Vienna to address the rinderpest problem in Europe through a uniform international approach (Internationale Conferenz zur Erzielung eines gleichförmigen Vorgehens gegen die Rinderpest).[46] The conference was attended by representatives of Belgium, Germany, France, Great Britain, Italy, Russia, Switzerland, Turkey, Serbia, the Romanian princely states, and delegates from Hungary. The national delegations consisted mainly of veterinarian experts, many of whom had been active at the previous international congresses, as well as of representatives from the ministries of the interior, agriculture, commerce, and railways.[47] Not surprisingly, the conference centred on the question of 'defending' Europe from, in the eyes of experts from Western and Central European countries, the incoming waves of rinderpest from Russia, Turkey, Serbia and Romania.[48] The meeting took up the contentious points about rinderpest that had already been discussed at the previous international veterinary congresses. The aim was to organise an international agreement to facilitate the trade in livestock by establishing a common regulatory framework, focusing on the idea of reforming national veterinary services to the same scientific and administrative standards and creating a system of mandatory exchange of information between states.[49]

To prepare for the conference, the Austrian organisers established a detailed questionnaire with 66 points outlining what they considered the most significant problems in rinderpest control. The conference discussed and voted on the points. They included questions on the nature of the rinderpest outbreaks especially in Russia and numerous issues about the effectiveness of preventive and repressive measures in dealing with the disease—on transport, disinfection, culling, sanitary borderlines, trustworthiness of health certificates, quarantine periods, animal species affected, contaminated products and transport material, hygienic removal of carcasses, implementation of import bans, the usage of products from culled animals, the issue of compensation of livestock owners, and more details.[50] The Central and Western European experts regarded the efforts of rinderpest vaccination conducted by researchers in Russia as highly dangerous.[51] For most participants, the usage of veterinary sanitary measures such as trading bans, border regulations with quarantines and a strict policy of culling seemed a better option for protecting their national economic interests, despite the trade-hampering effects and even though opinions were very divided on the dimension of these measures.[52] The conference finally agreed on "Principles for an International Regulation for the Eradication of Rinderpest".[53] This, as such, was a novelty and a success. Barely a decade earlier, international talks on epizootics control had first begun with meetings of experts; now there was a set of regulatory principles on rinderpest control, focusing on sanitary veterinary measures, agreed upon by an intergovernmental meeting. The very detailed regulatory principles summed up many of the points discussed at the previous international congresses. They highlighted a core feature of future international anti-disease work—the necessity of a systematic and regular exchange of information on the disease situation by modern communication technologies. Every participating veterinary administration should register disease cases in a weekly bulletin, and they should communicate outbreaks of rinderpest as fast as possible via telegraph to the other national authorities.[54] The conference participants also agreed on the principle of compensating livestock owners for their losses due to rinderpest, in order to reduce illegal trade practices.[55] Further points included scientific and technical regulations on disinfection procedures in cross-border animal transportations, on the border control stations, the health certificates for animals, the sanitary conditions concerning the export of animal products and the control of export markets, as well as the culling of animals and the disposal of carcasses.[56] The British delegate Alexander Williams, secretary of the veterinary department of the Privy Council, expressed doubts on the measures, arguing with regard to the British free trade tradition that the orientation towards border-focused strategies as the central means of disease containment was insufficient for adequate disease control.[57]

The regulatory principles of the conference were published in German and French; they were also intended as a basis for further intergovernmental negotiations on rinderpest control.[58] In 1878, experts from the field of phytopathology initiated an international convention to contain phylloxera, an insect that affects vines, by methods of disinfection and certification in an effort to create a safe transfer of vine plants.[59] This international convention served as a precedent for similar efforts in other fields. In the case of epizootics, nevertheless, the existence of a set of international principles on rinderpest control established by an intergovernmental conference marked real progress in the eyes of many experts. A number of European governments used the principles as a basis for their measures and could successfully contain rinderpest in the next decades.[60] The principles of rinderpest control also began to have an impact outside Europe. A decade later, James Law, a U.S.-American epizootics specialist and associate of Gamgee, reviewed the regulations of the conference at Vienna and highlighted the tentative efforts of international coordination as a model to improve U.S. veterinary regulations.[61] Law shared the view that sanitary border inspections as only measures were insufficient and hampering trade relations. In his opinion, the state veterinary services, if they existed at all, needed to overcome national mindsets and adopt more flexible attitudes, for example through a transregional orientation that improved the communication about the epizootics situation between neighbouring states.[62] The Vienna rinderpest conference of 1872, however, not only gave inputs by establishing a set of internationally discussed principles; it also had an impact on further governmental activities. In 1876, the Swiss government launched a similar initiative for international coordination to deal with the spread of bovine pleuropneumonia. The disease was a particularly invisible border-crossing problem, since its symptoms were not easily noticeable and it had a longer period of incubation. Borders alone certainly did not suffice to keep it successfully out of a territory. Other governments, however, did not follow the initiative—bovine pleuropneumonia had, after all, not yet resulted in similar destruction as rinderpest. The initiative was quickly given up.[63]

Towards an international convention on epizootics containment?

Encouraged by the Vienna rinderpest conference of 1872 and the success of the international convention on phylloxera in 1881, for many veterinary experts active in veterinary internationalism a formal international agreement in the field of animal diseases seemed to be getting more likely. In 1883, the veterinary community met again for the Fourth International Veterinary Congress, this time taking place in Brussels on the occasion of the 50-year anniversary of the Royal Veterinary School of Cureghem.[64] Now, 20 years after the first congress of 1863, the agenda of veterinary internationalism had expanded considerably. Due to the recent scientific-medical developments, the academic discipline and the practices of veterinary medicine was undergoing a fundamental change, which was reflected in the programme of the congress. Questions of veterinary public health, especially issues of food safety and animal tuberculosis, were increasingly prominent topics of discussion. Still, the agenda of the congress of Brussels showed that, while there was an ever-increasing array of themes, the original questions discussed for 20 years about containing the cross-border spread of epizootics were still present.[65] Auguste Zundel, the head veterinarian of Alsace-Lorraine, outlined in his contribution at the Brussels congress how the problem of an internationally coordinated system of animal health was inextricably linked with the ongoing process of building up professional, academically oriented veterinary services and raising the social status of veterinarians.[66] Having experienced shifting national borders and the associated professional and administrative problems as an Alsatian, Zundel was obviously interested in initiating stronger cross-border cooperation with regard to epizootics control. He outlined several pillars of national and international anti-disease regulation, documentation, and cooperation: direct communication of the animal disease situation through telegraphy would provide, in his opinion, an additional, faster level of mutual notification of concerned other national and regional administrations—an idea which had also featured in the discussions of the conference of 1872. In addition, in Zundel's view, neighbouring provinces from different countries needed to establish their own regional cross-border contact lines of communication.[67] Zundel expressed his wish of a close and mutually benefitting cooperation especially through the organisation of a new international bulletin which provided a certain level of guarantees for trading partners:

> The bulletin therefore gives the greatest mutual guarantee for trade and international relations, and is called upon to give mutual trust between the nations which will have to deal with each other in the traffic of domestic animals. May nations understand each other better in the area of epizootics than they generally do in the area of politics! Perhaps one step will lead to another![68]

The Brussels congress passed a resolution which demanded again the establishment of regular state-organised veterinary services and an international convention on the control of animal diseases. It should include the mutual notification of outbreaks of epizootics, a permanent system of information transfer about the disease situation, a common regulation of the methods of epizootics containment, and certificates of origin for traded animals issued by state veterinary services. Also the plan of an international epizootics bulletin was included.[69] Further points of the recommendations referred to the control of bovine pleuropneumonia, the improvement of veterinary education, veterinary pharmaceutics, and the expert's assessments concerning the question of the sale of the meat of tuberculous cows.[70]

The following Fifth International Veterinary Congress in 1889 took place, like many other international events, in context of the World's Fair in Paris. The international attendance by now was enlarged beyond European countries, with veterinarian experts coming from Japan, Egypt, and the United States. Instructed not to join the congress, German veterinary experts were officially absent.[71] The number of enrolled participants had risen to over 600.[72] Bovine tuberculosis—and especially the question of the sale of meat of suspected cattle—was again one of the main discussed points. On the question of an international agreement concerning epizootics, some experts began to show impatience. Hubertus Thomassen, a veterinarian from Utrecht who delivered a report on the question of international cooperation concerning animal disease control, wondered what success the past international gatherings had actually achieved and what sense it made to aim for an international convention.[73] Thomassen outlined that at least the previous international talks had led to a generally heightened awareness of the problem area from the part of governments, and the principles discussed in 1872 had in parts created inputs for a number of bilateral conventions.[74] Other experts joined in expressing their dissatisfaction with the stagnating situation in the question of an international agreement. Camille Leblanc, veterinary officer of the Seine department in France, complained of a general atmosphere of unwillingness and hesitation concerning any binding commitment.[75] It seemed, Leblanc pointed out, that to keep the status quo of a patchwork of different systems of national veterinary control, with its many loopholes and the great flexibility to shut down all borders at any time, was comfortable for governments.[76] He underlined that the use of sanitary arguments as pretexts for protectionism was an all-too-common practice, and the information provided by national veterinary bulletins was not sufficient to be of any value for establishing trustworthiness in the respective veterinary systems.[77] The propositions of the congress finally recommended an extended control of tuberculosis in animals and, to improve food safety, the complete removal of the meat from infected animals from the market. Further points referred to the control of bovine pleuropneumonia, to principles of compensation for livestock owners in case of culling, and questions of meat control. Concerning suggestions for international veterinary sanitary cooperation, the congress repeated and updated the points already made at previous congresses and especially highlighted the necessity of a closer cooperation in issues of sanitary border control.[78]

Most governments did not feel prompted by the veterinary congresses to initiate concrete steps towards more international cooperation. Leblanc's own activities already after the 1883 congress in Brussels revealed a different way of advancing international veterinary commitment. The idea of an international bulletin on the epizootics situation recommended in the resolutions of the congress in Brussels did not induce a reaction from the governments. A group of French veterinarians therefore started to approach the issue as a collaborative project. Together with his colleague Paul Cagny, Leblanc founded an international bulletin, the *Bulletin Trimestriel International des Epizooties*. They planned to publish the bulletin regularly in French and German and include information on the respective national situation of animal diseases in different countries. In 1883, a first edition appeared.[79] In the preface, Leblanc and Cagny described the project as a non-official initiative. They emphasised how in the recent years the concept of contagion through pathogens had transformed the perception of epizootics as cross-border challenges and thereby even more enhanced the practical significance of international cooperation. After the Brussels congress, Leblanc and Cagny had thought that it was the time for more activities of networking beyond national borders, but nothing had happened and the situation could, in their view, no longer wait for governmental initiatives:

> Given the way the recommendation was adopted, we thought that the psychological moment had arrived to move forward with its implementation and that it was not indispensable to wait until European governments decided whether to convene an International Animal Disease Commission... in order to speed up the implementation of the recommendation, we have considered that it might be useful to publish from now on a similar bulletin, which will not have an official character, but on the contrary will be a private initiative of the European veterinarians.[80]

Leblanc and Cagny organised their initiative for an international epizootics bulletin by sending an invitation to correspondents of the Société Centrale de la Médecine Vétérinaire, a French association of veterinary professionals, and to their personal contacts in the veterinary community. They asked all experts to send documents with disease-related information about their respective national or regional field of work and to distribute their agenda to other veterinarians charged with such work.[81] For the first edition, they were already able to assemble a considerable network of veterinarians from European countries to collaborate, including a veterinarian colleague from New York. Most were officials in the state veterinary services, and many of them renowned academics.[82] In order to ensure a large readership and contribution, it was important for Leblanc and Cagny that the new international bulletin was distributed for free to cooperators and important journals. The project was, as they outlined, financed with the support of Henri Bouley, professor at the Alfort veterinary school and one of the main French authorities on veterinary legislation, and the French academic journal *Recueil de Médecine Vétérinaire*, of which Bouley was editor-in-chief.[83] Leblanc and Cagny saw the project as a work in progress: they stated that the bulletin was still very incomplete and incoherent, but it was essential to start with some form of project, however rudimentary.[84] The bulletin thus contained preliminary statistics for a number of diseases in several European countries. Further, Leblanc and Cagny announced extensive future plans, which would indeed have pre-empted governmental duties in the field. They intended to publish annual maps indicating the prevalence and movements of epizootics, as well as focus more on veterinary public health with information about meat hygiene and the sanitary situation of abattoirs. They admitted that accomplishing a level of uniformity in disease descriptions of the statistics was indeed a major problem.[85] Accordingly, they published a detailed list of principles on what information was needed for the statistics of an international bulletin, and how to organise this information.[86] These principles included a list of localities in which animal infections appeared, of diverse relevant information on the respective animals (numbers of livestock slaughtered, sick, suspected, inoculated, and dead after or before inoculation) and on the loss of livestock through diseases. For highly contagious diseases of potentially epizootic proportions, such as bovine pleuropneumonia or rinderpest, a special version of the international bulletin was planned in order to deliver information quickly enough to take appropriate measures.[87]

After the first edition in 1883, however, Cagny and Leblanc stopped the project. Six years later, at the International Veterinary Congress in Paris in 1889, Leblanc stated that the main difficulty in organising the international bulletin had been the resistance of governments.[88] Since 1872, he explained, governments had insisted that international initiatives in veterinary matters would be taken care of by means of a future international agreement. Leblanc characterised this attitude of governments as disadvantageous and disappointing because of the lack of progress in the matter. He also pointed out that, not surprisingly, many governments actively obscured the situation of animal diseases in their countries. A culture of covering up, he outlined, seemed the far easier and cheaper option for governments and livestock-industry lobbyists than any sustainable cooperation across borders.[89] The project of the international trimestrial bulletin and Leblanc's assessment outlined a number of fundamental problems of the international activities against animal diseases as well as against globalised threats in general. He described two groups with specific and not always overlapping interests, the veterinary experts and the governments, though most experts, including himself, held official positions in the state veterinary services, and many experts represented the interests of their governments at international meetings. In this interpretation, for reasons of commercial interests and national prestige, many governments hesitated to promote a culture of international exchange of information on the animal disease situation. Leblanc's and Cagny's project not only highlighted the question of who had the agency and legitimacy to initiate international activities against animal diseases. It also pointed to the possibilities of advancing international cooperation through the networking of experts or at least putting pressure on governments to take appropriate action, even though the international trimestrial bulletin of epizootics was short-lived.

In the 1890s, after three decades of debates at international meetings, a set of possible international approaches for addressing cross-border animal disease problems was at hand, updated in the resolutions of several international congresses. In 1895, participants at the Sixth International Veterinary Congress in Berne made new efforts to advance these ideas.[90] Also Emmanuel Leclainche, a veterinarian from Toulouse who would after the First World War become the dominant figure in veterinary internationalism as the first firector of the OIE, attended the congress.[91] In the programme, the debate on international coordination appeared on several levels, with reports on the question of uniform sanitary services, the issue of an international bulletin, and in the form of a debate on internationally uniform nomenclatures of anatomy.[92] Younger veterinarian experts such as the Hungarian Ferenc Hutyra were impatient with the decade-long standstill of the question of international coordination of disease control measures.[93] In his contribution at the congress, he underscored the supposedly non-political role of veterinarian experts and their independence from national questions of trade policies.[94] Hutyra suggested again the idea of addressing directly the governments, this time the Swiss one as the host country of the congress, for taking up the initiative on intergovernmental cooperation. In order to realise this project more quickly, he called for a uniform regulation of the veterinary services and the livestock trade, with templates for all national bulletins on animal diseases and for health certificates for traded animals.[95] The congress participants followed his suggestions in the resolutions and recommended for the governments to establish an international intelligence service as well as an international bulletin on animal diseases. They expressed the wish that the Swiss government initiated not only the establishment of an international information service on animal diseases together with an international bulletin, but also an international conference for the establishment of an international sanitary convention on veterinary matters.[96] This was a step forward from the part of the experts to call for future governmental commitment. Further recommendations of the congress included veterinary recommendations on a number of specific disease issues, among them the usage of new diagnostic tools such as mallein (for glanders) and tuberculin (for tuberculosis) as well as the demand for the introduction of obligatory meat inspection services in all countries represented at the congress.[97]

The following years, however, still saw no international diplomatic conference to advance an intergovernmental agreement on joint approaches of epizootics containment in Europe. In southern Africa, several colonial governments discussed the question of intergovernmental cooperation to handle the ongoing massive rinderpest crisis. When in 1896 the devastating African rinderpest panzootic reached the Cape Colony south of the Molopo river, leaving a trace of carcasses of cattle and wild animals, the governments of the region organised hastily a series of conferences to cooperate on its containment. In April 1896, veterinarian experts and representatives from several affected states and colonial territories came together in Mafeking in the Cape Colony (today Mahikeng, South Africa); already in late August, a second conference followed in Vryburg in the Cape Colony.[98] In August 1897, a third meeting took place in Pretoria in Transvaal, with participants from the Orange Free State, the Cape Colony, Natal, the South African Republic (Transvaal) as well as Portuguese and German representatives. The aim of this International Congress on the Study of Rinderpest, and the Means of Combating this Epizootic in Southern Africa (Congrès International pour l'Etude de la Peste Bovine, et des Moyens de Combattre cette Epizootie dans l'Afrique Méridionale) was to discuss the ongoing experiments with vaccination as well as to draft a coordinated approach of the governments.[99] The veterinary sanitary methods applied by the colonial authorities, mainly building fences as sanitary barriers, establishing rigorous quarantines and implementing stamping-out, contributed to conflicts between the colonisers and the African population, also because the measures went along with racial discrimination.[100] At the congress, there were doubts among the participants as to whether the European sanitary methods to contain rinderpest were suitable for African geographical and ecological conditions; even those used in India did not seem to work properly in southern Africa.[101] The experiments of producing immunisation conducted by several teams of researchers seemed a promising approach, but even though some of the new vaccination types contributed to the containment of the disease, the experimentations and their results were controversial.[102] The Polish-French researcher Jan Danysz of the Pasteur Institute in Paris had developed the Pretoria congress programme with the support of the Transvaal government, and he had also included the idea of a central international rinderpest station or bureau.[103] The international office should perform the tasks of a clearing office, centralising all information on the extension of the rinderpest outbreaks, and collect and distribute statistical material and the results of the different experimental efforts. Further, according to this plan, it should publish the information and coordinate the treatment of rinderpest in all participating countries, as well as continue with organising experimental research.[104] A permanent international commission would advise and inform governments and organise the bureau's activities.[105] Indeed, Danysz aimed at "a complete entente between the interested governments"[106] to address rinderpest. These aims were not achieved, but the congress showed how significant it was to think beyond the conventional Europe-centred ideas of veterinary internationalism and veterinary approaches. Four further conferences took place until 1909, addressing the rinderpest problem in Southern Africa.[107] The vaccination efforts eventually helped to control rinderpest outbreaks, but the disease remained highly problematic for decades, especially in East Africa.[108]

By the time of the Seventh International Veterinary Congress in Baden-Baden (Germany) in 1899, many European veterinarian experts were disappointed by the stagnation on the question of an international agreement. Hutyra presented a detailed draft of an international veterinary convention text which suggested international standards of veterinary services and uniform procedures of the transport of animals across borders. It also included a template for the international assessment of information on animal diseases.[109] He pronounced his frustration with the passiveness of governments after decade-long initiatives, requests and recommendations from the part of veterinarians.[110] It seemed unnecessary, Hutyra observed, to point out the significance of the matter for the economy and for public health—the non-veterinarian authorities seemed to be disinclined to move towards international cooperation in the matter.[111] In his view, one of the most problematic and dangerous new issues which urgently needed to be addressed internationally was animal tuberculosis.[112] In Europe, he explained, there was by now a confusing variety of testing methods, different sorts of tuberculin, and interpretations and consequences of test results. Bilateral arrangements were, he thought, insufficient from both an economic and a veterinarian point of view, since they could not provide appropriate orientation for farmers and administrations.[113] Hutyra therefore proposed a resolution to the congress asking the host regional government of Baden to request the German imperial government to convene an international conference on the subject of veterinary regulation.[114] The Swiss federal head of epizootics control, Denys Potterat, explained that his government had not advanced the recommended project of an intergovernmental conference on joint measures of epizootics control further because it had expected it to fail.[115] Moreover, the veterinary experts attending the meeting were far from unanimous. Notions of what constituted a 'civilised' veterinary organisation began to show their underlying ambiguities, and national rivalries among experts were more openly expressed. The director of the veterinary school of Hanover, Karl Dammann, discarded the idea of international arrangements as impracticable and failing when it came to difficult detailed questions.[116] Eventually, the Baden-Baden congress participants voted for a proposition of recommending international coordination between governments and the adoption of uniform principles of a scientific veterinary system, but with a limitation:

> Considering the difference in the economic development and the transport conditions, as well as the dissimilitude of veterinary organisation in the different States, they do not think the moment has yet come for laying down definite principles for an international agreement.[117]

The congress thus moved a step away from the resolutions of previous congresses, which had outlined to the governments the need for a new international convention. The resolutions also included, among other points, principles for the containment of foot-and-mouth disease, as well as recommendations for effective meat inspection concerning tuberculous animals.[118]

At the next two International Veterinary Congresses, taking place in 1905 in Budapest and in 1909 at The Hague, the topic of an international convention or other projects of international coordination were not any more at the foreground. The congresses had developed into huge platforms for the professional exchange of an enlarged scientific community. At the Budapest congress, the agenda of specialised scientific-veterinary topics reflected the central role of new research in bacteriology, virology and immunology for veterinary medicine, as well as the increasing relevance of veterinary public health. Furthermore, a section on tropical veterinary medicine was included.[119] The congress also regulated and modernised its own structures and organisational procedures. After the event, a Permanent Committee for International Veterinary Congresses was established, which, together with the executive committee of the respective host country, organised the topics of the congress series by contacting veterinary schools of the world. August Lydtin, former head veterinarian of Baden, served as chairman, Ferenc Hutyra as vice-president, and fifteen veterinarians as committee members.[120] A secretariat of the Permanent Committee was established in the Dutch Ministry of Agriculture at The Hague to organise the future international congresses.[121] The following congress at The Hague in 1909 discussed a multiplicity of specialised topics, including food safety, veterinary public health, and tropical veterinary medicine.[122] Veterinary congress internationalism was now, in the years before the First World War, at a peak, with a closely networked scientific community, and consolidated as a transnational platform of scientific exchange. But the participants had largely abandoned the original aims highlighted at the congresses of the late 19th century, namely the establishment of a permanent international veterinary cooperation in the form of an official international convention and an international bureau addressing the cross-border control of epizootics through the systematic exchange and transfer of information.

Veterinary internationalism in August 1914

The Tenth International Veterinary Congress was scheduled to take place in London in August 1914 to celebrate the 50th anniversary of the first congress in Hamburg in 1863; originally it had already been planned for 1913. The programme, published in scientific journals, reflected the enormous enlargement of the veterinary scientific community and of its scientific spectrum in the previous decades.[123] Despite the recent establishment of international offices in related fields, the International Institute of Agriculture (Institut International d'Agriculture, IIA) in Rome in 1905 and the International Public Health Office (Office International d'Hygiène Publique, OIHP) in Paris in 1907, the idea of a similar institution for epizootics control was not dominant in discussions. The congress in London opened in early August 1914, at first in defiance of the day by day increasing international tensions and the resulting worries and practical problems. The simultaneity of the outbreak of the war and the opening of the congress throws a spotlight on the international veterinary community at the exact moment of the collapse of internationalism in its 19th-century organisational forms. Not surprisingly, it was from the start a politicised event, presenting for its participants a clash between their national interests and their ideal of universal scientific work. The congress demonstrated the fragility of the underlying rationale of scientific and veterinary internationalism, based on the idea of regular international communication and exchange of the scientific community. To hold a huge international congress in Britain was in any case, as it had turned out, a challenge for the veterinary scientific community. The British Foreign Office supported the congress by sending official invitations, but the government did not contribute financially to the congress, as for instance the Dutch government had done for the previous one.[124] The British organising committee therefore had to raise huge sums through member subscriptions and voluntary contributions.[125] In the weeks before the start of the congress, the organising committee had watched the development of the international crisis in summer 1914 with growing concerns.[126] During the last days of July, several congress participants with an official position were prevented from travelling to London; others were already underway and journeys could only be cancelled with difficulty. By early August, participants from numerous countries had registered in London.[127] The German delegation's passage was called off at the last minute, although several German veterinarians arrived nevertheless, among them August Lydtin, the president of the Permanent Committee for International Veterinary Congresses. Some Germans left again almost immediately.[128] On August 1, two days before the congress opened, *The Times* reported on "The European Theatre of War" in a new column entitled "The War Day by Day".[129] The same day, the Permanent Committee met for the usual preparatory session presided by Lydtin, albeit in a reduced circle of participants. The committee members tried to demonstrate somehow an attitude of normalcy, discussing the agenda of the congress and details of the statutes, such as official languages and the question of the location for the next event.[130] At the same time, the British organising committee debated whether to postpone the event, deciding eventually to open it.[131]

The president of the congress, Sir John McFadyean, principal of the Royal Veterinary College in London and one of Britain's renowned veterinary pathologists, officially opened the event with around 300 participants on the morning of August 3.[132] Despite the worrying situation, McFadyean tried to not give up the solemn celebration of the international veterinary community. Welcoming the foreign participants on behalf of the United Kingdom's veterinary profession, he gave an overview on the history and significance of the international veterinary congresses.[133] McFadyean emphasised that veterinarians needed to exert their influence to advance the prevention of animal diseases more insistently in the future. In most cases, prevention could only be effectuated by coordinated activities between the state authorities and expert forces.[134] He could not refrain from making a remark about the lack of initiative on the part of government officials of several countries: "It would, unfortunately, be easy to show that incalculable sums of money would have been saved by some countries had their Governments paid earlier heed to the resolutions passed at some of these congresses".[135] McFadyean, though very worried, still expressed to the audience his hopes that the congress would be "a mark of the advancement of veterinary knowledge and the veterinary profession throughout the civilised world".[136] The veterinary-scientific part of the congress was planned to start the following day. The remaining official delegations wished the congress success in their welcome speeches.[137]

The same day, Germany declared war on France, and in the night German troops began the invasion of Belgium. On August 4, the participants of the veterinary congress could read in *The Times* that "the position of Europe is one of breathless anticipation of the beginning of hostilities on a large scale".[138] Many left London immediately. The organisers tried, often unsuccessfully, to obtain written summaries of the announced papers and presentations.[139] On the same morning of August 4, the first general meeting of the congress was held.[140] It was not completely clear which participants or delegations had already left. McFadyean summarised the dilemma of the congress organisers and expressed his worries:

> When the Congress was opened yesterday we all felt that a cloud of anxiety was hovering over it, and today the cloud has become much darker. Without any circumlocution I may say that, in the unexpected and absolutely unparalleled circumstances, it seems to me that we must now very seriously consider whether we should endeavour to conduct our deliberations according to the programme, or whether, on the contrary, it would not be better to bring the Congress immediately to a close.[141]

Internationalism—here in the form of professional veterinary exchange of a scientific community—was in a state of dissolution, and its objectives were becoming seemingly irrelevant, as McFadyean explained:

> The first painful fact which we have to recognise is that the Congress, as a Congress, has ceased to be international… the public mind is absolutely filled with affairs of immensely greater importance than those with which the Congress is concerned.[142]

McFadyean continued that many of the British attendees had to answer the order of mobilisation. Meeting again in the afternoon, the remaining participants of the congress accepted the proposal to close the meeting. Concerns about the future of veterinary internationalism were expressed.[143] McFadyean declared the congress closed. The same day, in the evening, Britain declared war on Germany. Despite the terminated congress, in 1915 the organisers published the congress volumes.[144] As they stated, they were able to include most written reports scheduled for the general and sectional meetings of participants.[145]

In its aftermath, some German veterinarians used the congress event for nationalist purposes, which showed how quickly the international veterinary community could be divided. Still on August 6, the *Berliner Tierärztliche Wochenschrift* (BTW), an important veterinary journal in the German language, celebrated the jubilee event ("Dem Jubiläumskongreß!")[146] taking place five decades after the first congress in Hamburg; it had obviously been too late to stop the printing. The editor of the journal, the Berlin veterinary professor Reinhold Schmaltz, took the opportunity in the preface to express his greetings to British veterinarians and praise international cooperation.[147] Schmaltz emphasised the close scientific ties between Germany and Britain as being mirrored by the congresses, with the Scotsman Gamgee having initiated the first meeting in 1863.[148] Schmaltz's appraisal was followed by a short summary of the past congresses by Erwin Wehrle, director of the veterinary department in the Kaiserliches Gesundheitsamt (Imperial Health Office), who would in 1933/34 act as president of the OIE.[149] On August 13, the next issue of the journal referred to the congress "in arms" ("Jubiläumskongreß in Waffen").[150] International meetings and friendships were deceitful, Schmaltz outlined in its preface, since they veiled the true intentions of those involved. He expressed his relief that the German delegation had not arrived in London in time.[151] In the following years, the BTW was used as a forum for war propaganda, and Schmaltz tried to initiate a German-dominated veterinary community of the Central Powers. In 1917, he started to lobby for the dissolution of the Permanent Committee for International Veterinary Congresses at The Hague with the aim of establishing a new, Central-European organisation to be founded after the war, with German as its academic language.[152] Schmaltz began to pressure the committee's president August Lydtin and its vice-president Ferenc Hutyra for the dissolution of the Permanent Committee. Both declared themselves, in accordance with the statutes, not entitled to take such a step.[153] Schmaltz still continued to express himself against veterinary internationalism.[154] The war thus resulted in a reversal of discourses concerning veterinary internationalism—previously, the international contacts and networking activities had been considered by many experts a sine qua non of being part of the circle of 'civilised' states and their scientific community. Not surprisingly, in Schmaltz's version of international veterinary organisation, scientific and veterinary arguments were marginalised. At least in parts, Schmaltz even succeeded in promoting his ideas—the Permanent Committee for

International Veterinary Congresses did not dissolve but ceased its activities for many years when its chairman August Lydtin died in 1917; the series of International Veterinary Congresses was only resumed during the 1930s.

The war presented veterinarians and researchers with the needs of adjustments to military objectives and thus transformed national veterinary systems and practices. The military organisation of all warring parties was heavily dependent on animals for transport and provisioning. The principles of fighting epizootics with measures related to territorial organisation, based on the idea of defence from the exterior and, if possible, prophylaxis within the interior, remained valid also in times of war, but the focus shifted in order to secure the army supplies. The Balkan wars and the outbreak of rinderpest in their context had demonstrated that warfare not only changed borderlines but also disrupted territorial sanitary systems.[155] The sanitary controls at borderlines, one of the main strategies against the spread of epizootic diseases in Europe, became in context of the war often difficult to be maintained. Both state and military borders shifted, sometimes vanished, and were crossed by the movements of different armies, transport systems, and trails of refugees with their animals.[156] New allocations of territories changed former regional organisations of disease control. In regions without frontlines, there were often no veterinarians available to ensure the basics of animal disease control. Veterinarians underscored their military-related professional significance and prestige as members of their respective national army veterinary services. The armies established veterinary hospitals, recovery depots for horses, and bacteriological laboratories.[157] The military commands directed the use of horses and livestock as war material.[158] In the case of the German army, the military organised a new territorial classification of animal disease incidences, dividing them into the spatial categories of home, occupation, or operation.[159] The area of operation had to be protected to prevent military movements from being disrupted by the occurrence of epizootics. The emphasis, above all, had to be on preventive measures, conducted through elaborate military logistic systems.[160] In case of an outbreak of an epizootic, the most radical measure, stamping-out, was difficult to apply on a large scale near frontlines. Preventive measures included maintaining the hygiene of stables and the extended disinfection of means of transports, stables and carcasses. Since adequate chemicals were rare, in the case of highly contagious animal diseases, disinfection meant burning the carcasses and used material as much as possible.[161] Diseases of horses, for example glanders or mange, were among the central problems. Glanders, often fatal in human beings, was a serious threat not only to the animals but to the soldiers.[162] The First World War also created unexpected new problems of animal health. Heavy artillery and bombing began to increase the numbers of dead and wounded animals. Veterinarians began to debate new disease problems such as those relating to gas poisoning of horses. The development of gas masks for military animals symbolised the enormous relevance of animals for globalised warfare, and at the same time its terrible absurdity.[163]

Notes

1 For the history of the telegraph in global contexts: Roland Wenzlhuemer, *Connecting the Nineteenth-Century World: The Telegraph and Globalization* (Cambridge, New York: Cambridge University Press, 2015).

2 About the developments of internationalism in the 1860s: Madeleine Herren, *Internationale Organisationen seit 1865. Eine Globalgeschichte der internationalen Ordnung* (Darmstadt: Wissenschaftliche Buchgesellschaft, 2009), 18–25; Madeleine Herren, 'International Organizations, 1865–1945', in *The Oxford Handbook of International Organizations*, ed. Jacob Katz Cogan, Ian Hurd, and Ian Johnstone (Oxford: Oxford University Press, 2016), 94–5.

3 Herren, *Internationale Organisationen seit 1865*, 18–9.

4 Herren, *Internationale Organisationen seit 1865*, 16–7.

5 Herren, *Internationale Organisationen seit 1865*, 17–27; Herren, 'International Organizations, 1865–1945', 95–7.

6 Herren, *Internationale Organisationen seit 1865*, 19–22.

7 For an overview: Neville M. Goodman, *International Health Organizations and Their Work* (Edinburgh: Churchill Livingstone, 1971), 318–26. For a detailed analysis: Mark Harrison, *Contagion: How Commerce Has Spread Disease* (New Haven: Yale University Press, 2012), 139–73.

8 On the international sanitary conferences: Valeska Huber, '"The Unification of the Globe by Disease"? The International Sanitary Conferences on Cholera, 1851–1894', *Historical Journal* 49, no. 2 (2006): 453–76; Mark Harrison, 'Disease, Diplomacy and International Commerce: The Origins of International Sanitary Regulation in the Nineteenth Century', *Journal of Global History* 1, no. 2 (2006): 197–217.

9 Adrien Proust, *La Défense de l'Europe contre le Choléra* (Paris: G. Masson, 1892); Adrien Proust, *La Défense de l'Europe contre la Peste et la Conférence de Venise de 1897* (Paris: Masson, 1897) (translation: CK).

10 There exist several summary accounts on the series of International Veterinary Congresses from veterinary experts, some of them having attended a number of the events: Erwin Wehrle, 'Geschichte der internationalen tierärztlichen Kongresse', *Berliner Tierärztliche Wochenschrift* 30, no. 32 (1914): 567–81; Reinhold Schmaltz, *Entwicklungsgeschichte des tierärztlichen Berufes und Standes in Deutschland* (Berlin: Schoetz, 1936), 341–55; Gottlieb Flückiger, 'Bisherige Bemühungen zur internationalen Bekämpfung tierischer Seuchen', *Schweizer Archiv für Tierheilkunde* 96, no. 1 (1954): 12–5; Emmanuel Leclainche, *Histoire Illustrée de la Médecine Vétérinaire*, vol. 2 (Paris: A. Michel, 1955), 183–209.

11 John Gamgee (1831–1894), Scottish veterinarian and inventor of refrigeration techniques, founded the New Edinburgh Veterinary College in 1857. About Gamgee: Clive A. Spinage, *Cattle Plague: A History* (Dordrecht and London: Kluwer, 2003), 222–39; Pamela Hunter, *Veterinary Medicine: A Guide to Historical Sources* (Aldershot: Ashgate, 2004), 215–16; Ruth D'Arcy Thompson, *The Remarkable Gamgees: A Story of Achievement* (Edinburgh: Ramsay Head Press, 1974); Michael Worboys, *Spreading Germs: Disease Theories and Medical Practice in Britain, 1865–1900* (Cambridge and New York: Cambridge University Press, 2000), 48–9.

12 Michael Worboys, 'Germ Theories of Disease and British Veterinary Medicine, 1860–1890', *Medical History* 35 (1991): 310–11; Joanna Swabe, *Animals, Disease, and Human Society. Human-Animal Relations and the Rise of Veterinary Medicine* (London and New York: Routledge, 1999), 73.

13 Swabe, *Animals, Disease, and Human Society*, 73.

14 Circular, in: Eduard Hering and Wilhelm Probstmayr, eds., *Amtlicher Bericht über die internationale thierärztliche Versammlung zu Hamburg am 14–18.*

Juli 1863 (Stuttgart: Ebner & Seubert, 1864), 1–2. Eduard Hering (1799–1881) was veterinary professor at the Tierarzneischule (veterinary school) Stuttgart (Württemberg); see: Joachim Bossneck, 'Eduard von Hering', in *Neue Deutsche Biographie*, vol. 8 (Berlin: Duncker & Humblot, 1969), 617.

15 Circular, in: Hering and Probstmayr, *Bericht internationale thierärztliche Versammlung 1863*, 1.

16 Circular, in: Hering and Probstmayr, *Bericht internationale thierärztliche Versammlung 1863*, 1.

17 Hermann von Nathusius and Anton Krocker, *Landwirthschaftliches Thier-Album in Photographien: Die Internationale Landwirthschaftliche Ausstellung in Hamburg, 1863* (Berlin: Wiegand und Hempel, 1866). About the congress see also: Harrison, *Contagion*, 215; Swabe, *Animals, Disease, and Human Society*, 73–4; Amanda Kay McVety, *The Rinderpest Campaigns: A Virus, Its Vaccines, and Global Development in the Twentieth Century* (Cambridge: Cambridge University Press, 2018), 20-1.

18 Hering and Probstmayr, *Bericht internationale thierärztliche Versammlung 1863*, 2–3.

19 Hering and Probstmayr, *Bericht internationale thierärztliche Versammlung 1863*, 2–3.

20 Hering and Probstmayr, *Bericht internationale thierärztliche Versammlung 1863*, 3.

21 Hering and Probstmayr, *Bericht internationale thierärztliche Versammlung 1863*, 4–5.

22 Hering and Probstmayr, *Bericht internationale thierärztliche Versammlung 1863*, 5–8.

23 Röll, in: Hering and Probstmayr, *Bericht internationale thierärztliche Versammlung 1863*, 7. About the discussion on quarantine regulations see also: Harrison, *Contagion*, 215.

24 Hering and Probstmayr, *Bericht internationale thierärztliche Versammlung 1863*, 5–13.

25 For example Müller, in: Hering and Probstmayr, *Bericht internationale thierärztliche Versammlung 1863*, 8.

26 Hering and Probstmayr, *Bericht internationale thierärztliche Versammlung 1863*, 14–22.

27 Fuchs, in: Hering and Probstmayr, *Bericht internationale thierärztliche Versammlung 1863*, 22. Christian Joseph (Josef) Fuchs (1801–1871) was since the 1860s one of the main organisers of the veterinary services in Baden and author of numerous veterinary studies; see: Friedrich von Weech, ed., 'Christian Josef Fuchs', in *Badische Biographien, 1. Lieferung* (Heidelberg: Bassermann, 1875), 270–1.

28 Hering and Probstmayr, *Bericht internationale thierärztliche Versammlung 1863*, 23–31.

29 Hering and Probstmayr, *Bericht internationale thierärztliche Versammlung 1863*, 30–1.

30 Closing speech of Hering, in: Hering and Probstmayr, *Bericht internationale thierärztliche Versammlung 1863*, 41–3.

31 About the congress and the context see also: Harrison, *Contagion*, 215–16.

32 George Fleming, *A Manual of Veterinary Sanitary Science and Police*, vol. 1 (London: Chapman & Hall, 1875), xiv; Swabe, *Animals, Disease, and Human Society*, 74–6; Harrison, *Contagion*, 216–20.

33 Angela von den Driesch and Peter Joris, *Geschichte der Tiermedizin. 5000 Jahre Tierheilkunde* (Stuttgart: Callwey, 2002), 168; Patrick Zylberman, 'Making Food Safety an Issue: Internationalized Food Politics and French Public Health from the 1870s to the Present', *Medical History*. 48, no. 1 (2004): 4–5; Harrison, *Contagion*, 215.

34 Circular of Invitation, in: Moritz Friedrich Röll and Leopold Forster, *Amtlicher Bericht über den zweiten internationalen Congress von Thierärzten zu Wien am 21.–17. August 1865* (Wien: Wilh. Braumüller, 1865), 2–3.

35 Verzeichniss (sic) der Herren Mitglieder, in: Röll and Forster, *Amtlicher Bericht Congress 1865*, 5–8.

36 John Gamgee, *The Cattle Plague. With Official Reports of the International Veterinary Congresses, Held in Hamburg, 1863, and in Vienna, 1865* (London: Hardwicke, 1866), 530–1.

37 Röll and Forster, *Amtlicher Bericht Congress 1865*, 163.

38 Beschlüsse des Congresses: Schlussfassung über die Principien einer rationellen Hunde-Ordnung, in: Röll and Forster, *Amtlicher Bericht Congress 1865*, 164.

39 Verzeichnis der Herren Mitglieder, in: Rudolf Zangger, ed., *Amtlicher Bericht über den dritten internationalen Congress von Thierärzten zu Zürich am 2.–7. September 1867* (Zürich: Schabelitz, 1869), 3–6.

40 Zangger, *Amtlicher Bericht Congress 1867*, 7 (translation: CK). Carl Schenk (1823–1895) was a Swiss liberal politician and Federal Councillor. Christoph Zürcher, 'Schenk, Carl', Historisches Lexikon der Schweiz, accessed October 17, 2020, https://hls-dhs-dss.ch/articles/004636/2011-07-12/.

41 Zangger, in: Zangger, *Amtlicher Bericht Congress 1867*, 8. The Swiss veterinarian and politician Rudolf Zangger (1826–1882), was from 1856 to 1882 director of the veterinary school (Tierarzneischule) in Zurich. About Zangger see: Markus Bürgi, 'Zangger, Rudolf', Historisches Lexikon der Schweiz, accessed April 1, 2021, https://hls-dhs-dss.ch/de/articles/003731/2014-02-03/.

42 Zangger, *Amtlicher Bericht Congress 1867*, 106.

43 Zangger, *Amtlicher Bericht Congress 1867*, 107.

44 Zangger, *Amtlicher Bericht Congress 1867*, 108.

45 Jean Wehenkel, ed., *Compte-Rendu du IVe Congrès International de Médecine Vétérinaire, Bruxelles 1883* (Bruxelles: Brogniez et Vande Weghe, 1884), 392.

46 Johann von Chlumecký, ed., *Verhandlungen der internationalen Conferenz zur Erzielung eines gleichförmigen Vorgehens gegen die Rinderpest, gehalten vom 16. März bis 6. April 1872 in Wien* (Wien: Kaiserlich-Königliche Hof- und Staatsdruckerei, 1872), 12. The editor of the conference report and head of the general sessions, the politician and lawyer Johann von Chlumecký (1834–1924), was Austrian Minister of Agriculture. About the conference see also: Zylberman, 'Making Food Safety an Issue', 4–6; McVety, *The Rinderpest Campaigns*, 21–2.

47 Mitglieder der Conferenz, in: Chlumecký, *Verhandlungen der Conferenz 1872*, 12.

48 See also: Zylberman, 'Making Food Safety an Issue', 4.

49 See also: Zylberman, 'Making Food Safety an Issue', 5–6.

50 Questionnaire, in: Chlumecký, *Verhandlungen der Conferenz 1872*, 5–11.

51 Chlumecký, *Verhandlungen der Conferenz 1872*, 23. About the Russian experiments on rinderpest immunisation see: Wilhelm Dieckerhoff, *Geschichte der Rinderpest und ihrer Literatur. Beitrag zur Geschichte der vergleichenden Pathologie* (Berlin: T. C. F. Enslin, 1890), 227–8; Spinage, *Cattle Plague*, 410–15.

52 In detail about the different argumentations: Zylberman, 'Making Food Safety an Issue', 5.

53 Grundsätze für ein internationales Regulativ zur Tilgung der Rinderpest, in: Chlumecký, *Verhandlungen der Conferenz 1872*, 87–91 (translation: CK). On the regulation see also: McVety, *The Rinderpest Campaigns*, 21–2.

54 Grundsätze für ein internationales Regulativ zur Tilgung der Rinderpest, II, in: Chlumecký, *Verhandlungen der Conferenz 1872*, 87. See also: Dieckerhoff, *Geschichte der Rinderpest*, 242; McVety, *The Rinderpest Campaigns*, 21–2.

55 Grundsätze für ein internationales Regulativ zur Tilgung der Rinderpest, III, in: Chlumecký, *Verhandlungen der Conferenz 1872*, 87. See also: McVety, *The Rinderpest Campaigns*, 21–2.

56 Grundsätze für ein internationales Regulativ zur Tilgung der Rinderpest, IV-VI, in: Chlumecký, *Verhandlungen der Conferenz 1872*, 87–9. See also: McVety, *The Rinderpest Campaigns*, 21–2.

57 Anhang zu den Conferenz-Verhandlungen, Exposé des englischen Delegierten Herrn Dr. Williams, in: Chlumecký, *Verhandlungen der Conferenz 1872*, 91–5. See also: McVety, *The Rinderpest Campaigns*, 22.

58 Österreich-Ungarn, K. K. Ackerbau-Ministerium, *Bericht über die Thätigkeit des K. K. Ackerbau-Ministeriums in der Zeit vom 1. Jänner 1869 bis 30. Juni 1874. Erster Theil* (Wien: Faesy und Frick, 1874), 185.

59 The phylloxera convention was initiated by a private assembly of wine-growers who successfully approached the governments for international cooperation. About the convention of 1878 (modified 1881) and its assessment as a "prototype of international cooperation among agriculturalists" see: Frederick William Kohlmeyer, *The Movement toward International Cooperation in Food and Agriculture: Background of the FAO of the UN* (PhD thesis, University of Minnesota, 1954), 46–8 (citation: 46); Madeleine Herren, *Hintertüren zur Macht: Internationalismus und modernisierungsorientierte Aussenpolitik in Belgien, der Schweiz und den USA, 1865–1914* (München: Oldenbourg, 2000), 253–6.

60 Ferenc Hutyra, 'Bericht/Entwurf einer internationalen Veterinär-Konvention', in *Siebenter Internationaler Tierärztlicher Kongress: Baden-Baden, 7.–12. August 1899*, vol. 1 (Baden-Baden: Hofdruckerei Ernst Kölblin, 1899), 6.

61 James Law, 'Report of the International Veterinary Congress of Brussels', ed. United States Bureau of Animal Industry, *Annual Report of the Bureau of Animal Industry*, 1885, 321–70, here 340–1. James Law (1838–1921), a born Scotsman, served as first veterinary professor at Cornell University, then as dean of the New York State College of Veterinary Medicine. About Law: James H. Steele, 'Zoonoses 1879–1983: An Update of James Law's Report on Diseases of Animals', *Veterinary Heritage* 23, no. 1 (2000): 14–20.

62 Law, 'Report of the International Veterinary Congress of Brussels', 328.

63 Law, 'Report of the International Veterinary Congress of Brussels', 326. About the Swiss governmental initiative for international cooperation against bovine pleuropneumonia: (Auguste) Zundel, 'Rapport: Organisation de Service Vétérinaire', in *Compte-Rendu du IVe Congrès International de Médecine Vétérinaire, Bruxelles 1883*, ed. Jean Wehenkel (Bruxelles: Brogniez et Vande Weghe, 1884), 108.

64 Wehenkel, *Compte-Rendu congrès 1883*, VII.

65 Law, 'Report of the International Veterinary Congress of Brussels', 326.

66 Zundel, 'Rapport: Organisation de Service Vétérinaire', 101.

67 Zundel, 'Rapport: Organisation de Service Vétérinaire', 143–4.

68 Zundel, 'Rapport: Organisation de Service Vétérinaire', 144 (translation: CK).

69 Propositions admises par le Congrès, in: Wehenkel, *Compte-Rendu congrès 1883*, 623–4.

70 Propositions admises par le Congrès, in: Wehenkel, *Compte-Rendu congrès 1883*, 624–30.

71 As reported by: Schmaltz, *Entwicklungsgeschichte des tierärztlichen Berufes und Standes in Deutschland*, 345.

72 Liste des Membres du Congrès, in: Saturnin Arloing, ed., *Ve Congrès International de Médecine Vétérinaire, tenu à Paris du 2 au 8 Septembre 1889: Recueil de Rapports* (Paris: Asselin et Houzeau, 1889), 9–30.

73 (M. H. J. P.) Thomassen, 'Rapport: Service Sanitaire International', in *Ve Congrès International de Médecine Vétérinaire, tenu à Paris du 2 au 8 Septembre 1889: Recueil de Rapports*, ed. Saturnin Arloing (Paris: Asselin et Houzeau, 1889), 89.

74 Thomassen, 'Rapport: Service Sanitaire International', 90.

75 About the veterinarian and researcher Camille Leblanc (1827–1903) see: Comité des Travaux Historiques et Scientifiques (Ecole Nationale des Chartes),

'Louis Camille Leblanc', Annuaire Prosopographique: La France Savante. Comité des Travaux Historiques et Scientifiques (Ecole Nationale des Chartes), accessed March 18, 2021, https://cths.fr/an/savant.php?id=4034.

76 Camille Leblanc, 'Rapport: Service Sanitaire International', in *Ve Congrès International de Médecine Vétérinaire, tenu à Paris du 2 au 8 Septembre 1889: Recueil de Rapports*, ed. Saturnin Arloing (Paris: Asselin et Houzeau, 1889), 146.

77 Leblanc, 'Rapport: Service Sanitaire International', 146.

78 Propositions adoptées par le Congrès, in: Arloing, *Recueil des rapports congrès 1889*, 601–3.

79 Camille Leblanc and Paul Cagny, eds., *Bulletin Trimestriel International des Epizooties*, 1er et 2e trimestres 1883 (Paris: Recueil de Médecine Vétérinaire, 1883).

80 Leblanc and Cagny, *Bulletin Trimestriel International des Epizooties*, 6 (translation: CK).

81 Leblanc and Cagny, *Bulletin Trimestriel International des Epizooties*, 6–7.

82 The contributors were: Röll (Austria), Bagge (Denmark), Fisher (Luxembourg), Lima (Portugal), Liautard (USA), Wehenkel (Belgium), Zundel (Alsace-Lorraine), Berdez (Switzerland), Cope (England), Goering (Bavaria), Lindqvist (Sweden), Lydtin (Baden), Müller (Berlin), Neshan (Bulgaria), Potterat (Switzerland), Schmoulewitsch (Russia), Siedamgrotski (Russia), Wirtz (Netherlands). Collaborateurs étrangers, in: Leblanc and Cagny, *Bulletin Trimestriel International des Epizooties*, 13–14.

83 About the veterinarian and researcher Henri Bouley (1814–1885): François Vallat, *Les Bœufs Malades de la Peste. La Peste Bovine en France et en Europe, XVIIIe-XIXe siècle* (Rennes: Presses Universitaires de Rennes, 2009), 130–1.

84 Leblanc and Cagny, *Bulletin Trimestriel International des Epizooties*, 6–7.

85 Leblanc and Cagny, *Bulletin Trimestriel International des Epizooties*, 7.

86 Leblanc and Cagny, *Bulletin Trimestriel International des Epizooties*, 14.

87 Leblanc and Cagny, *Bulletin Trimestriel International des Epizooties*, 8.

88 Leblanc, 'Rapport: Service Sanitaire International', 149.

89 Leblanc, 'Rapport: Service Sanitaire International', 149.

90 About the congress see: B. Hörning and R. Fankhauser, 'Vor 100 Jahren: 6. Internationaler Tierärztlicher Kongress, September 1895 in Bern', *Schweizer Archiv fur Tierheilkunde* 138, no. 1 (1996): 8–11.

91 Leclainche, *Histoire Illustrée de la Médecine Vétérinaire*, 2: 188. In detail about Leclainche's work in veterinary internationalism see Chapters 4 and 6.

92 The question of uniform nomenclatures had already been discussed at earlier congresses, such as in Paris 1889, see: Leblanc, 'Rapport: Service Sanitaire International', 147.

93 Ferenc Hutyra (also: Franz von Hutyra, 1860–1934) was from 1899 to 1931 rector of the Hungarian Royal Veterinary College in Budapest. Together with József Marek, he authored a standard work on diseases of domestic animals: Ferenc Hutyra and József Marek, *Spezielle Pathologie und Therapie der Haustiere* (Jena: Fischer, 1905).

94 Ferenc Hutyra, 'Internationale Tierseuchen-Polizei', in *VIe Congrès International de Médecine Vétérinaire, tenu à Berne, du 16 au 21 Septembre 1895: Rapports et Compte-Rendu*, ed. Emile Noyer (Bern: Stämpfli, 1896), 175.

95 Hutyra, 'Internationale Tierseuchen-Polizei', 188–9.

96 Résolutions votées par le Congrès, in: Emile Noyer, ed., *VIe Congrès International de Médecine Vétérinaire, tenu à Berne, du 16 au 21 Septembre 1895: Rapports et Compte-Rendu* (Bern: Stämpfli, 1896), 853 (point 1).

97 Résolutions votées par le Congrès, in: Noyer, *Rapports et Compte-Rendu Congrès 1895*, 853–7.

98 Spinage, *Cattle Plague*, 536, 541.
99 C. E. Schutte, ed., *Comptes-Rendus du Congrès International tenu à Prétoria du 2 au 13 août 1897 pour l'Etude de la Peste Bovine et des Moyens de Combattre cette Epizootie dans l'Afrique Méridionale* (Pretoria: Staatsdrukkerij van de Zuid-Afrikaansche Republiek, 1898), 5–6. Translation of conference title: CK.
100 McVety, *The Rinderpest Campaigns*, 26–7.
101 Schutte, *Comptes-Rendus du Congrès International tenu à Prétoria 1897*, 7, 51–2.
102 Schutte, *Comptes-Rendus du Congrès International tenu à Prétoria 1897*, esp. 18–49. See also: Spinage, *Cattle Plague*, 553. On the immunisation experiments also: McVety, *The Rinderpest Campaigns*, 27–30.
103 Schutte, *Comptes-Rendus du Congrès International tenu à Prétoria 1897*, 6. About the Polish bacteriologist and biologist Jan (Jean) Danysz (1860–1928) and his work on rinderpest inoculation in Southern Africa together with the Belgian physician and bacteriologist Jules Bordet (1870–1961) see Marc Mammerickx, 'La Peste Bovine, Jules Bordet et le Centre Sérumigène de Cureghem', *Annales de Médecine Vétérinaire* 147 (2003): 197–200.
104 Schutte, *Comptes-Rendus du Congrès International tenu à Prétoria 1897*, 6.
105 Danysz, in: Schutte, *Comptes-Rendus du Congrès International tenu à Prétoria 1897*, 66.
106 Danysz, in: Schutte, *Comptes-Rendus du Congrès International tenu à Prétoria 1897*, 67 (translation: CK).
107 Klemens Wedekind, *Impfe und herrsche: Veterinärmedizinisches Wissen und Herrschaft im kolonialen Namibia 1887–1929* (Göttingen: Vandenhoeck & Ruprecht, 2021), 155.
108 McVety, *The Rinderpest Campaigns*, 29–30.
109 Hutyra, 'Bericht/Entwurf internationale Veterinär-Konvention', 14–32.
110 Hutyra, 'Bericht/Entwurf internationale Veterinär-Konvention', 5.
111 Hutyra, 'Bericht/Entwurf internationale Veterinär-Konvention', 9.
112 Hutyra, 'Bericht/Entwurf internationale Veterinär-Konvention', 7–8.
113 Hutyra, 'Bericht/Entwurf internationale Veterinär-Konvention', 7, 12.
114 Hutyra, 'Bericht/Entwurf internationale Veterinär-Konvention', 13.
115 Lettre du Président du Comité d'Organisation du VIme Congrès [Denys Potterat] au Président du Celui du VIIme [August Lydtin], concernant le Sort des Resolutions prises à Berne sur l'Elaboration d'un Bulletin Sanitaire International et sur une Convention réglant le Trafic International d'Animaux Domestiques, in: *Siebenter Internationaler Tierärztlicher Kongress: Baden-Baden, 7.–12. August 1899*, vol. 1 (Baden-Baden: Hofdruckerei Ernst Kölblin, 1899), 160–1. Denys Potterat (1843–1913) was from 1890 to 1910 head of the Swiss Federal epizootics control administration; see: Sarah Brian Scherer, 'Potterat, Denys', Historisches Lexikon der Schweiz, accessed March 19, 2020, http://mobile.hls-dhs-dss.ch/m.php?article=D17449.php.
116 Dammann, in: *Siebenter Internationaler Tierärztlicher Kongress: Baden-Baden, 7.–12. August 1899*, vol. 2 (Baden-Baden: Hofdruckerei Ernst Kölblin, 1899), 17–19. Karl Dammann (1839–1914) was for more than three decades director of the veterinary school in Hanover. See: Wilhelm Pschorr, 'Karl Dammann', in *Neue Deutsche Biographie*, vol. 3 (Berlin: Duncker & Humblot, 1957), 500–1.
117 Resolutions of the Congress, in: *Siebenter Internationaler Tierärztlicher Kongress*, 1899, 2: 518. See also: Verhandlungen, in: *Siebenter Internationaler Tierärztlicher Kongress*, 1899, 2: 27.

118 Resolutions of the Congress, in: *Siebenter Internationaler Tierärztlicher Kongress*, 1899, 2: 518–23.
119 István Rátz, ed., *Bericht über den VIII. internationalen tierärztlichen Kongress in Budapest 1905*, vol. 1, 3 vols (Budapest: Pátria, 1906).
120 A. de Jong, 'Préface', in *Travaux Du IXe Congrès International de Médecine Vétérinaire à La Haye, 13–19 Septembre 1909*, vol. 3, 3 vols (Leiden: Brill, 1910), 1–4. Members: 17; Réunion et Séance de la Commission Permanente des Congrès Internationaux de Médecine Vétérinaire, Lyon, October 25, 1912. ADLC: Unions, 429QO/658 (Affaires Diverses: Commission Permanente des Congrès Intern. de Médecine Vétérinaire); see also: Reinhold Schmaltz, Bisherige Entwicklung und künftige Gestaltung der internationalen tierärztlichen Kongresse, in: Rátz, *Bericht Kongress 1905*, 1: 612–22. August Lydtin (1834–1917) was head of the Baden (Germany) veterinary services until 1895 and author of numerous veterinarian works. See his obituary: 'Personalien – Geheimer Oberregierungsrat Dr. August Lydtin', *Schweizer Archiv fur Tierheilkunde* 59 (1917): 642–3.
121 For information about the Bureau and the Permanent Committee for International Veterinary Congresses at The Hague see the entry in the database LONSEA, beginning with the League's *Handbook of International Organization* of 1925: LONSEA, 'Permanent Committee for International Veterinary Congresses', accessed March 19, 2020, http://www.lonsea.de/pub/org/107.
122 *Travaux du IXe Congrès International de Médecine Vétérinaire à La Haye, 13–19 Septembre 1909*, vol. 2, 3 vols (Leiden: Brill, 1910), here section V, 2.
123 Stewart Stockman, ed., *Tenth International Veterinary Congress*, vol. 1 (London: J. Bale Sons & Danielsson, 1915), 146–8. About the congress see also: Bruce Vivash Jones, 'Unfortunate Timing...', *Veterinary Record* 174, no. 25 (2014): 627–9.
124 Stockman, *Tenth International Veterinary Congress*, 1: V–VI.
125 Stockman, *Tenth International Veterinary Congress*, 1: 179. See also about the funding: Jones, 'Unfortunate Timing…', 627.
126 Stockman, *Tenth International Veterinary Congress*, 1: V.
127 Stockman, *Tenth International Veterinary Congress*, 1: VI. Also: Schmaltz, *Entwicklungsgeschichte des tierärztlichen Berufes und Standes in Deutschland*, 352–3.
128 For a contemporary German perspective: Kurt Schern, 'Bericht aus London', *Berliner Tierärztliche Wochenschrift* 30, no. 38 (1914): 677–8.
129 'The War Day by Day', *The Times*, August 1, 1914.
130 Permanent Committee of the International Veterinary Congresses, Minutes of the Meeting at London, August 1, 1914. Hoover Institution Archives: Germany, DKZ, Box 307 (Tierärzte-Kongress, 1863–1942).
131 Stockman, *Tenth International Veterinary Congress*, 1: VI; Jones, 'Unfortunate Timing…', 628–9.
132 About John McFadyean (1853–1941) see the biography: Iain Pattison, *John McFadyean: A Great British Veterinarian* (London and New York: J. A. Allen, 1981).
133 McFadyean, in: Stockman, *Tenth International Veterinary Congress*, 1: 171–6; see also: Jones, 'Unfortunate Timing…', 629.
134 McFadyean, in: Stockman, *Tenth International Veterinary Congress*, 1: 175.
135 McFadyean, in: Stockman, *Tenth International Veterinary Congress*, 1: 176.
136 McFadyean, in: Stockman, *Tenth International Veterinary Congress*, 1: 176.
137 Portugal, Italy, Japan, Belgium, the Netherlands, Denmark, Russia, Chile, Romania, Argentine, Sweden, Egypt, USA. Official Opening Meeting, in: Stockman, *Tenth International Veterinary Congress*, 1: 170–89. See also: Jones, 'Unfortunate Timing…', 629.

138 'The War Day by Day', *The Times*, August 4, 1914.
139 Stockman, *Tenth International Veterinary Congress*, 1: V.
140 Stockman, *Tenth International Veterinary Congress*, 1: 186. On the meeting and McFadyean's efforts see also: Jones, 'Unfortunate Timing…', 629.
141 McFadyean, in: Stockman, *Tenth International Veterinary Congress*, 1: 186.
142 McFadyean, in: Stockman, *Tenth International Veterinary Congress*, 1: 186.
143 Stockman, *Tenth International Veterinary Congress*, 1: 190–9, here 196. See also about the closure of the congress: Jones, 'Unfortunate Timing…', 628–9.
144 McFadyean, in: Stockman, *Tenth International Veterinary Congress*, 1: 198–9.
145 Stockman, *Tenth International Veterinary Congress*, 1: 193.
146 Reinhold Schmaltz, 'Dem Jubiläumskongreß!', *Berliner Tierärztliche Wochenschrift* 30, no. 32 (1914): 565–6.
147 The *Berliner Tierärztliche Wochenschrift* was founded and edited by Reinhold Schmaltz (1860–1945), professor of veterinary medicine at the University of Berlin. About Schmaltz: Martin Fritz Brumme, *Lebenswege eines wilhelminischen Bürgers. Biographische und anatomiegeschichtliche Untersuchungen zu Reinhold Schmaltz (1860–1945)* (Freie Universität Berlin: Dissertation, 1992).
148 Schmaltz, 'Dem Jubiläumskongreß!', 566.
149 Wehrle, 'Geschichte der internationalen tierärztlichen Kongresse'. About Erwin Wehrle (1868–1936), director of the veterinary department in the German Kaiserliche Gesundheitsamt/since 1918 Reichsgesundheitsamt (Reich Health Office) and in 1933/34 president of the OIE committee, see the obituary: 'Erwin Wehrle', *Tierärztliche Rundschau* 42 (1936): 662.
150 Reinhold Schmaltz, 'Jubiläumskongreß in Waffen', *Berliner Tierärztliche Wochenschrift* 30, no. 33 (1914): 597–8.
151 Schmaltz, 'Jubiläumskongreß in Waffen', 598.
152 Schmaltz, in: *Berliner Tierärztliche Wochenschrift* 33, no. 9 (1917): 97.
153 Hutyra and Lydtin, in: *Berliner Tierärztliche Wochenschrift* 33, no. 13 (1917): 156.
154 Reinhold Schmaltz, 'Zur tierärztlichen Internationale', *Berliner Tierärztliche Wochenschrift* 33, no. 17 (1917): 202.
155 Ferenc Hutyra and József Marek, *Die Orientalische Rinderpest. Mit besonderer Berücksichtigung der klinischen und anatomischen Merkmale und der Differentialdiagnose* (Jena: Fischer, 1916), III–V.
156 Hutyra and Marek, *Die Orientalische Rinderpest*, III.
157 For Britain: John Moore, *Army Veterinary Service in War* (London: Brown, 1921), 30–5, 37–8. For the British case see also Hunter, *Veterinary Medicine: A Guide to Historical Sources*, 352.
158 A contemporary assessment about veterinary work in the First World War, from a German perspective, is: M. Reuter, 'Tierseuchen und sporadische Tierkrankheiten im Kriege. Deren Auftreten und Bekämpfung bis zum Stande vom 1. Mai 1916', in *Ergebnisse der Hygiene, Bakteriologie, Immunitätsforschung und experimentellen Therapie*, ed. W. Weichardt, vol. 2 (Berlin: Springer, 1917), 668–747.
159 Reuter, 'Tierseuchen und sporadische Tierkrankheiten im Kriege', 672–3.
160 Reuter, 'Tierseuchen und sporadische Tierkrankheiten im Kriege', 675–80.
161 Reuter, 'Tierseuchen und sporadische Tierkrankheiten im Kriege', 674.
162 Reuter, 'Tierseuchen und sporadische Tierkrankheiten im Kriege', 711–17.
163 For an overview on the use of animals in the First World War, including the development of gas masks for horses and carrier pigeons: Christoph Jahr and Stefan Kaufmann, 'Den Krieg führen: Organisation, Gewalt, Technik', in *Erster Weltkrieg: Kulturwissenschaftliches Handbuch*, ed. Niels Werber, Stefan Kaufmann, and Lars Koch (Stuttgart: Metzler, 2014), 206–7.

4 Restarting veterinary internationalism after the First World War

In the years after the First World War, outbreaks of animal diseases aggravated the already difficult worldwide economic situation. In Europe, it was the spread of foot-and-mouth disease and the appearance of rinderpest that prompted new international activities in the early 1920s. In 1914, half a century of activities in congress internationalism in veterinary matters had come to an abrupt end. After the war, internationalism was undergoing a profound transformation. The new post-war international system combined old and new elements of international organisation and considerably enlarged its circle of participants.[1] The founding of the League of Nations and the International Labour Organization (ILO) marked a new beginning in efforts of international cooperation. Some international institutions, such as the OIHP in Paris and the IIA in Rome, had survived the war. Others, especially non-governmental initiatives such as the series of international veterinary congresses and its Permanent Committee for International Veterinary Congresses at The Hague, did not show activities since the war, and it was unclear when or whether they would resume their work. Already in 1921, the Conférence Internationale pour l'Etude des Epizooties (International Conference for the Study of Epizootics) convened in Paris to discuss plans for a new international office on epizootics control, and in 1924 a number of governments agreed on founding the Office International des Epizooties (OIE) in Paris. The project of a new international office aimed to take up, stabilise and expand the existing concepts of veterinary internationalism—establishing a reliable and permanent system of information transfer, and creating a new international platform of exchange and future cooperation to address questions of animal disease control.

The discussions at the International Conference for the Study of Epizootics in Paris in 1921 showed the complexity of animal disease control in a global context. The tensions between the prevalent European focus of veterinary internationalism and the perceived global dimension of disease spread influenced the debates about the role and direction of the envisaged new office. Also the question how to incorporate issues of veterinary public health into the new projects of veterinary internationalism was controversial.

DOI: 10.4324/9781003033998-4

On the one hand, the conference discussions reflected more than before the awareness of the global dimension of the animal disease problem and the need to address this problem through the activities of veterinary internationalism. On the other hand, the conference debates showed that most concepts and practices of veterinary internationalism continued to be tailored to the requirements and concerns of European experts and governments. Although the global dimension of the spread of animal diseases was recognised as one of their main characteristics and problems, the animal disease concerns of non-European countries or of colonies were not central issues of veterinary internationalism in the early 1920s. Separate regional organisations beyond Europe did not yet exist or address the cross-border dimension of epizootics. The debates on a new beginning of international cooperation in epizootics control in the interwar years thus began with a dilemma: while contemporary experts active on international platforms discussed animal diseases in their complex and interdisciplinary character and with view to their spread also in global contexts, they did only in limited ways conceptualise and mobilise pathways of international cooperation beyond the concerns of European governments and experts.

Post-war Europe and globally connected animal diseases

The movements of people and animals in the context of the war activities—continuing in several parts of the world—and their aftermath enhanced the spread of contagious diseases of humans and animals. Severe epidemics accompanied the trails of refugees or the travels of soldiers, also on a global level. Especially the influenza pandemic of the years 1918–1920 caused millions of human fatalities around the world. In the U.S., epidemiologists and veterinarians discussed a zoonotic background of the disease when observing a simultaneous and highly lethal outbreak of an epizootic of influenza of hogs, but the issue was controversial among the scientists.[2] In countries that had been severely struck by the war, or in which fighting continued, the territorial borders, and therefore also the national or colonial veterinary systems, had shifted or were disrupted. The structures of public health and of veterinary control were often in a precarious state, and famines, malnutrition and diseases were frequent consequences. Outbreaks of epizootics, for example of foot-and-mouth disease and rinderpest, contributed to the economic misery of the population and to problems of food supply.

In the post-war discussions on the veterinary situation in both European and global contexts, two intertwined tendencies became apparent. First, the question of appropriate disease control measures took on a political dimension, continuing the use of arguments about the quality of veterinary and sanitary control to advance geopolitical interests. Second, the thematic area of epizootics was characterised at the same time clearly as a technical-scientific question. In international relations, this categorised the subject area as one for which veterinary experts could discuss and devise allegedly neutral and non-political solutions. The political dimension became visible in rivalries about the distribution of colonial power and the question of trustworthiness and legitimacy of the governments aiming for the control of the new mandate territories. In May 1919, the Belgian delegation at the Paris Peace Conferences highlighted their claims on parts of former German East Africa, the territories of Ruanda and Urundi. The British and American delegates expressed doubts about a territorial enlargement of Belgian colonial rule and objected to the attachment of the two regions to the colony of Belgian Congo. Among their arguments was the perceived danger of epizootics in Central Africa: they questioned Belgian colonial veterinary resources and competencies to provide adequate disease containment and expressed fears about severe consequences for the whole region. The Chief Veterinary Officer of the British East Africa Protectorate, Robert J. Stordy, launched a debate by sending letters to *The Times* about the issue. He argued that the potential threat of rinderpest was "perhaps the most immediate important consideration, because it directly involves the present welfare of the native population"[3] and could easily affect neighbouring British colonial territories. In the adjacent British protectorates of Uganda and East Africa, Stordy emphasised, there existed appropriate scientific laboratories to produce rinderpest serum for an undivided region. This would be of greatest benefit to the local population, which, Stordy thought, needed the guiding hand of competent (here meaning British) veterinary surgeons:

> I would plead the urgency of taking immediate steps to safeguard the animal industry of these primitive peoples, for, with that migration of wild game which accompanies the spread of animal plagues, the introduction of rinderpest to these areas must be looked upon as imminent.[4]

The Belgian Ministry of the Colonies replied by publishing a protest note, declaring the competence and the adequacy of the Belgian colonial veterinary administration in the event of a major epizootic outbreak.[5] At the end of May 1919, an Anglo-Belgian agreement divided the territory of former German East Africa, and in July 1919, the Mandates Commission of the Peace Conference confirmed the transformation of Ruanda-Urundi into B mandates under Belgian supervision. Veterinary argumentation had, in this case, not helped the British to assert their plans concerning African mandate territories. During the same period, the foreign ministries and experts often tried their best to characterise animal disease problems as purely technical and non-political issues. The definition of the subject area as a supposedly technical matter that was relevant to all countries had one decisive advantage—it facilitated post-war negotiations on the topic area with all parties involved, including former wartime enemies, and thus the creation of new international cooperation in the subject area. The critical situation of animal disease control in the context of a livestock sector devastated by war pushed governments and veterinary experts alike to increase cooperation and eventually made the cross-border problem of animal diseases a matter of international governmental activities.

The initiative to resume and strengthen international veterinary cooperation after the war came from the French government in context of efforts to protect the national agricultural economy. Similar to the situation in the 1860s, it was a response to immediate threats of epizootics affecting Western Europe. In 1920, Joseph-Honoré Ricard, an agricultural engineer and rural insurance expert who had just become Minister of Agriculture, started to focus his attention on the matter.[6] In a report published afterwards, Ricard named as his motivation for a new international initiative on animal diseases the concerns about the re-emerging waves of foot-and-mouth disease. Since the mid-19th century, the epizootic had repeatedly appeared in Europe, and in 1920 there was a particularly severe outbreak. Ricard visited affected regions in Switzerland during his holidays to study how the Swiss authorities were dealing with the disease, and he was impressed by what he described as a very thorough and comprehensive approach. An inquiry from the French Ministry of Agriculture into the methods of the Swiss veterinary strategies to control the epizootic followed.[7] As Ricard outlined in his report, the cooperation between French and Swiss experts revealed that the Swiss approach was based on the simultaneous application of diverse strategies. In addition to strict sanitary border controls and the immediate culling of diagnosed animals, the Swiss authorities had extended their measures to include vaccination efforts with the blood of healed animals, a controversial and not entirely convincing method. They also extended rigorous veterinary sanitary controls to other species of livestock.[8] As Ricard noted, a comprehensive national network of communication was the basis for an effective anti-disease alarm system. Local, cantonal and national administrations closely cooperated and informed each other about new cases. The Swiss public was regularly updated through the press, and local sanitary control stations even asked tourists to undergo meticulous disinfection procedures in the respective areas. These large-scale efforts had shown some amount of success, but the measures were still not sufficient to prevent outbreaks.[9] For Ricard, the experience showed the significance of a both flexible and comprehensive approach towards epizootics that needed to be conducted simultaneously on several local, regional and national levels, with a functioning system of information transfer as one of its central elements. In addition, he intended to complement the internal measures by new efforts of international coordination to prevent the spread across borders.

In summer 1920, Ricard was thinking about how to start and coordinate a new approach to address the problem of foot-and-mouth disease beyond national borders when alarming news arrived in Paris—an outbreak of rinderpest had been recorded in Belgium. As Ricard noted, the last notified outbreak of rinderpest in France had occurred in 1870/71, killing about 50,000 cattle through the disease and through culling.[10] Together with the wave of foot-and-mouth disease, the reappearance of rinderpest in Western Europe prompted the start of a new effort toward international cooperation on epizootics. The reactions to the Belgian rinderpest outbreak of

1920 displayed several perceptions on epizootics in their global dimension and their relevance for Europe. The outbreak—in peacetime—revived old anxieties connected with the disease since the 18th century. Further, it again showcased the interconnectedness of Europe with the world. It showed Europe as a central hub of a network of both globalised trade and connected disease flows contributing to disease transmission to other continents, and with serious deficiencies in terms of animal disease alert systems and prevention methods. Ricard reported that two days after the news had arrived in Paris in August 1920, a French technical delegate in Brussels reported to Paris that Belgium had already recorded 40 locations in the country where rinderpest was evident.[11] It was by now clear that the outbreak was without doubt a consequence of the worldwide trade in livestock which aimed to replenish the herds that had been decimated by the war. Ricard reported that rinderpest had been introduced into Belgium in July with zebus imported from British India.[12] The animals were in transit to Brazil and were put into interim stabling in Antwerp. Cattle from North America and from Germany stabled in Belgium spread the virus throughout the country.[13] In his report, Ricard presented the activities which followed like a lesson on how to organise an ad-hoc national defense program against an epizootic crisis approaching from abroad. The French government at once initiated the closure of the northern borders to transports of animals and animal products. Veterinarians were instructed to conduct strict surveillance of all animals, a three-sectorial zone of protection was established near the Belgian borders, and ports and stations were closed for animal transports. A special rinderpest commissioner was given the authority to order the culling of all suspected animals. The Ministry of Agriculture organised press announcements and public notices. This was especially necessary since the Belgian rinderpest outbreaks had already spurred a little voyeur tourism of "plague-watching" ("pour voir la peste").[14] Rinderpest vaccines existed since the early 20th century, with varying effectiveness,[15] but a stock of vaccines to create some form of a barrier of immunisation against the disease existed neither in France nor in Belgium. The French Ministry hastily organised a limited amount of vaccine via dispatch from the Serum Institute in Cairo, which specialised with a multinational team under British supervision in rinderpest research.[16] Belgian and French veterinary experts agreed on a close programme of cooperation to produce a suitable vaccine in an improvised special institute in Cureghem. Ricard was relieved that the spread of the disease could be stopped before it entered France, and after January 1921, no new outbreaks were recorded in Belgium.[17] The disease had been brought under control by a combination of strict sanitary measures including rigorous culling, the closure of borders for animals, the French scientific and organisational assistance and close collaboration of Belgian and French veterinarian experts both on site and in laboratory work, the veterinary monitoring of animals and scientific documentation of cases, the constant distribution of disease information, and international cooperation, such as with the Cairo Institute.[18]

The relief of the French and Belgian veterinary authorities on successfully containing the outbreak of rinderpest in Western Europe was short-lived since it mixed with the unpleasant recognition that, in the event of an epizootic of greater proportion spreading across borders, the preparedness of their veterinary systems had serious shortcomings. There had been no effective vaccine available in Europe at short notice, and while French and Belgian authorities cooperated well, especially through professional networks of experts, the episode revealed dangerous gaps in national veterinary systems and cross-border cooperation alike. Several bilateral treaties concerning animal trade regulations and control measures existed, but the improvised coordination between France and Belgium showed that there was decisively room for improvement. In particular, the old question of a regular and systematic transfer of information concerning disease outbreaks gained new topicality. After all, as Ricard noted, at the beginning of the rinderpest outbreak it had been more or less a coincidence that the French veterinary administration had been informed so early about the Belgian rinderpest cases, mainly through personal and professional contacts between veterinarian researchers in the two countries.[19] It was therefore unlikely that cooperation beyond borders could always be as fast and successful as that between the French and Belgian veterinarians.[20] According to Ricard and other French veterinary experts, cross-border cooperation on animal diseases—not only rinderpest, but all relevant contagious animal diseases—urgently needed a new organisational approach to prevent future damage. Ricard decided to follow the idea to advance a new "international entente" to address the problem of the spread of animal diseases in general, especially the lack of reliable and accessible information:

> From that moment on, it seemed to me indispensable to provoke an international entente and to establish a common information and documentation centre on the development and treatment of contagious animal diseases.... But how could this idea be realised? In the wake of concerns expressed in several states, could we risk pushing ahead with a scheme that aimed to cover not just one disease, but the ensemble of possible dangerous diseases?[21]

Aware of the fact that, at the international congresses before the war, veterinary experts had tried in vain to attract the interest of governments in this direction, the French organisers of the new international initiative were concerned about the possible pitfalls of the project. In Ricard's opinion, the concept of an international convention with a focus on regulating animal disease control procedures at borders and in ports to prevent the introduction of a limited number of specific diseases was insufficient. As he outlined, the initiative should, first, not concentrate on a specific disease but encompass a whole list of globally relevant contagious animal diseases with different patterns of spread and conditions of containment.[22] Second, he underlined the necessity of establishing a new international institution in order to build up a regular and comprehensive transfer of information and the documentation of the respective national veterinary situation. The new project would, as he stated, not intervene into national disease control systems, but link them through organised exchange and information transfer.[23] Role models were certainly the international institutions created before the First World War, such as the OIHP in Paris, the equivalent for human epidemics, or the IIA in Rome. At the same time, the activities of the new institution could be potentially linked to international programmes being developed in the context of the League of Nations, which started to build up new structures of international health with its Epidemics Commission for Eastern Europe (from April 1920) and its Provisional Health Committee (meeting for the first time in August 1921).[24] In the context of these new and large-scale international efforts, the timing seemed ideal for a new initiative to establish an institution addressing animal disease problems at the international level.

Paris 1921: the Conférence Internationale pour l'Etude des Epizooties

In the autumn of 1920, with the rinderpest outbreak in Belgium still causing concerns for Western European veterinary administrations, the French governmental and administrative apparatus began to take steps to convene the planned international conference on animal disease control. Emmanuel Leclainche, inspector general of the veterinary services at the French Ministry of Agriculture, played a key role in the conference project.[25] It was the starting point for Leclainche to become the predominant figure in veterinary internationalism for the next 25 years. The professor of veterinary medicine, almost 60 years old, had made a reputation for himself in the French and international scientific community through his work first in Toulouse and, since 1912, in leading positions at the Ministry of Agriculture in Paris; since 1913, he was the head of the research laboratory of the sanitary services at the Veterinary School in Alfort. In 1923, he would also become the inspector general of the French veterinary schools. In the early 1920s, therefore, he was regarded in France as "the undisputed head of the veterinary profession".[26] His qualifications as a researcher and veterinary organiser, together with his institutional and personal networks, made him one of the prominent figures in the international veterinary community of his time. Leclainche and the team of French organisers at the Ministry of Agriculture, together with the diplomats at the Ministry of Foreign Affairs, clearly aimed at a successful, presentable outcome of the international conference project. The strong governmental backing as well as the perceived urgency of the problem area indeed gave the veterinary and scientific experts organising and attending the conference a more influential voice than before the war. In November 1920, the French Ministry of Foreign Affairs sent an official circular to its diplomatic representatives with a description of the new project. The text indicated the hopes for a new conceptual and organisational start to veterinary internationalism:

> In the present circumstances, it is imperative that common health policing actions are undertaken towards contagious animal diseases. The economic imbalance caused by the war and the immense reconstruction efforts in the world show as a consequence the intensification of exchange. Animals are transported in large numbers and over long distances for supply and breeding purposes. Not only Europe and America participate in this, but the states and colonies in all parts of the world. Therefore, from now on, every country should not only take care of the sanitary situation of its immediate neighbours, but of that of the entire world.[27]

This approach to veterinary internationalism included several new conceptual elements. The text emphasised the significance of the global dimension of the problem area for the new project. A global scope implied a structural and conceptual enlargement of the European model of veterinary internationalism. It was, however, unclear how the plan of including all states and colonies into the new project was to be implemented. Furthermore, the initiative also aimed to include a focus on zoonotic diseases. This was an opportunity to integrate the problem of transmissible diseases between humans and animals on a permanent basis into international cooperation. Focusing on zoonotic diseases would present new options for combining international veterinary work across academic disciplines with issues of public health, beyond the dominant classification of animal diseases as matters of economic and agricultural policies. It would offer a chance to include in international discussion pressing concerns such as animal tuberculosis, brucellosis and animal influenza, and newly emerging diseases. The agenda also offered new options of possible cooperation, as well as potential rivalries, with other international organisations occupied with public health, such as the OIHP and the League of Nations. The aims of the conference organisers became clear in the summary presented at the end of the invitation letter:

> 1. Institutionalisation of an annual international conference on research into epizootic diseases and their prophylaxis. 2. Establishment of an international permanent office for epizootics with the following mission:
> a) to centralise all information on the occurrence of animal diseases and publish it rapidly;
> b) to collect all documents related to the study of these diseases; to follow and to initiate research on these diseases;
> c) to collect the results of different methods of prophylaxis (sanitary systems, methods of immunisation); and
> d) to prepare the work of the annual conference with preliminary studies on all issues of its agenda.[28]

Again, according to this plan, the organisers of the new project intended to include as many governments as possible. This was very necessary in view of the problematic animal disease situation in many parts of the world. But in the process of organising the conference, uncertainties emerged as to the desired worldwide scope of the new veterinary internationalism. The United States government, for example, expressed reservations about the project. The government, following its isolationist direction, expressed its regrets for not being able to attend the international meeting, giving financial reasons.[29] The United States with its important livestock sector was a member of the OIHP and the IIA, though not of the League of Nations. That the United States early on declared its non-participation in the new project on epizootics control was a serious setback from the start. Furthermore, considering the transboundary nature of the problem area and its importance for trade and transit policies, it made little sense to exclude the former war opponents from the new project. The organisers were clearly aiming for a broad international participation by highlighting the technical nature of the events. The invitation of smaller countries which were former war opponents, such as Bulgaria and Austria, did not raise many questions in the diplomatic correspondence. Sending an invitation to Constantinople seemed more difficult because of the ongoing political crisis in the Ottoman Empire, but an official invitation was transmitted in April 1921, just four weeks before the beginning of the conference.[30] The participation of Germany was another critical issue. It was justified by the French Ministry of Foreign Affairs again with the technical character of the conference.[31] Without Germany's participation, the organisers feared, the conference would not lead to practical results. After communicating with the governments in Rome, London and Brussels, three weeks before its opening, an invitation was sent to the German government asking to nominate veterinary experts for the conference, which was accepted.[32]

Another problematic issue was the participation of representatives of non-European countries as well as of colonies and mandate territories in the new project of veterinary internationalism. Most of them had so far not been represented in international veterinary work such as the pre-war congresses, though many faced immense problems of animal diseases. After the invitation to the conference arrived in London, the Foreign Office began to enquire about the future position of dominions and colonies in the project.[33] The French Ministry of Agriculture emphasised its aim to include representatives of British dominions and of India, especially with regard to their future participation in the envisaged new international office.[34] Furthermore, some non-European governments expressed difficulties with the European focus presented in the exposé. The government of Guatemala, for instance, pointed out a main impediment of many countries to more active participation in international veterinary work—their country provided neither a state-organised veterinary service with professional veterinarians nor any official veterinary statistics to take part in the envisaged regular transfer of disease information.[35] Already the process of organising the international meeting therefore presented several intrinsic and long-term problems of veterinary internationalism. Many countries could not send any veterinary experts to participate, either because there existed no veterinary sanitary system according to European standards, or their animal disease problems were different from those discussed (with other diseases or different species of livestock in the focus), the travel costs for sending representatives to Europe were too high, or colonial powers did not intend their participation. The invitation to the conference was therefore a diplomatically delicate action, since it disclosed incapacities in the field of national and colonial veterinary control. Even before its start, the envisaged global scope of the new project, in terms of the list of member states and of the focus on veterinary problem areas, became a difficult issue with regard to its realisation in practice.

On the morning of the first day of the international conference in May 1921, the organisers in Paris belatedly became aware of another problem: they had not prepared a statement on the question of the relationship between the League of Nations and the envisaged new international bureau of epizootics.[36] The staff of the Ministry of Foreign Affairs hastily prepared an evaluation of the issue to provide an answer based on the French line of policy. After all, the League intended to include multiple fields of technical and scientific cooperation. Article 24 of the Covenant suggested a new organisation of international offices in their relations with the League:

> There shall be placed under the direction of the League all international bureaux already established by general treaties if the parties to such treaties consent. All such international bureaux and all commissions for the regulation of matters of international interest hereafter constituted shall be placed under the direction of the League.[37]

The French government, however, did not want the planned new organisation to be based in Geneva. In addition, several important livestock-producing countries were not members of the League, again especially the United States; Argentina left the League in the same year, 1921. This fact could limit the future office in its scope of member states. The French argumentation, therefore, was to suggest an interpretation of Article 24 of the League Covenant as not binding for the discussion since it did not apply to non-League members. The new bureau should be planned without attachment to the League. Still, the French Ministry of Foreign Affairs saw the question as a delicate problem for French international engagement ("délicat pour la France").[38]

The conference finally convened in Paris on 25 May 1921.[39] The list of participants from 43 countries, including colonies and dominions, was impressive in comparison to pre-war veterinary internationalism. Many delegates were veterinary scientists or researchers attending as representatives of national sanitary institutions or ministries of agriculture, others were officials of agricultural ministries or diplomats. From overseas countries especially, it had been in some cases too costly to send a veterinary expert; therefore delegates were representatives of the respective legation without veterinary expertise. Still, it was a meeting of the international veterinary and scientific-medical elite, with many well-known researchers appearing on the list of participants. The French delegation was particularly numerous. Among them was the director of the Pasteur Institute, Emile Roux, and the deputy director Albert Calmette, as well as Henri Pottevin of the OIHP.[40] Henri Vallée was present, director of the national veterinary research laboratories at Alfort, who focused on foot-and-mouth disease vaccine research since the outbreaks of 1920.[41] Even the U.S. government sent its representative from the Bureau of Animal Industry (BAI) in England. Several Latin American and Carribean countries, among them Argentina as an important meat-producing country, sent delegations. From Asia, only Japan was represented by an official delegate. Apart from Morocco and Tunisia, represented by French experts, and South Africa with Arnold Theiler as delegate, no African country or colony was represented. Germany sent three veterinary experts (Erwin Wehrle, Robert von Ostertag and Friedrich Müssemeier); Great Britain and Canada were represented by Stewart Stockman.[42] The governments of Russia, China, and the Ottoman Empire were not represented, as well as most colonies and mandate states. Still, the meeting was larger, in terms of countries represented, than most international veterinary congresses before the war. The French officials, Ricard's successor as Minister of Agriculture Edmond Lefebvre du Preÿ and the senator Alfred Massé, carefully underlined in their opening speeches the sovereignty of all participating parties in matters of veterinary policies and the non-interventionist character of the meeting.[43] The situation of epizootics was described as serious and urgently requiring joint international activities. The Hungarian delegate Ferenc Hutyra, who during the 1890s had vehemently advocated for an international convention, showed himself intent on finally seeing results with regard to international veterinary cooperation. He recalled to the participants the first similar intergovernmental conference on rinderpest in Vienna in 1872, almost 50 years earlier, which he considered a starting point and symbolic of the indirect but important influence of expert authority over the decades. Hutyra outlined that many states had implemented parts of the resolutions elaborated at the 1872 conference; this had caused rinderpest to retreat from Europe for several decades. The new effort of a governmental implementation of international debates of experts, represented by this new conference, was in his eyes the most promising approach.[44]

In three plenary sessions and four commission meetings on specific animal disease concerns, the conference participants discussed in particular four topic areas. Firstly, they debated the recent outbreaks of rinderpest, foot-and-mouth disease, and dourine (a parasitic disease of horses) and the appropriate veterinary sanitary measures; secondly, the participants focused on the issue of how to organise a new system of information transfer, in particular through uniform animal disease reports and an international bulletin; thirdly, they looked at measures concerning the safe export of living animals and guarantees for importing countries; and fourthly, they discussed the creation of a new permanent international office dealing with the study and the containment of epizootics. The debates started with an exchange on the epizootic situation and the latest experiences with rinderpest, foot-and-mouth disease, and dourine.[45] The introductions into the topics were held by Emmanuel Leclainche in the name of the French delegation. In the case of rinderpest, he concentrated on the current epizootic situation as presenting particular problems to the newly created states in Eastern Europe, in his interpretation a result of shifting borders, dissolved health administrations, movements of people and the warfare of Bolshevik troops.[46] Leclainche's report highlighted the complex uncertainties surrounding the spread of epizootics, in the case of rinderpest also revolving around the question of which animals could transmit the disease and should therefore be included in the measures of culling. The role of pigs in the spread of the disease, for example, was not sufficiently clear.[47] Plus, he underlined the lack of mutual and reliable information on the disease situation which made combating cross-border epizootics a difficult task.[48] As it turned out in the discussion, rinderpest had appeared in Europe repeatedly during the last months, and not only in Belgium. In fact, regional conferences in Vienna and Kaunas with participants from several Central and Eastern European governments had in early 1921 tackled the question of rinderpest in Eastern Europe with regard to its cross-border spread.[49] The international conference on rinderpest in Vienna in January 1921 had discussed the issue of information transfer through a common intelligence system, aiming especially to control the disease in border regions. The participants had agreed on mutual short-term notifications about disease cases in the border regions via telegraph, the regular transfer of updated disease information, and on precautions in animal importations.[50] The conference at Paris in May 1921, similarly, debated new and old strategies of control, containment and prevention of rinderpest, such as the methods of quarantine, stamping-out, and the prophylaxis through methods of immunisation.[51] Since the pathways of disease transmission were not always clearly to be determined, there was the problem of which animals should be culled and which animal products destroyed[52]—a question that experts and veterinary authorities had discussed already for decades. Several experts, above all those with experience in colonial contexts, pointed out that one of the problems of European veterinarian systems in fighting epizootics was the lack of thinking beyond their conceptual framework.[53] In the case of foot-and-mouth disease, too, the discussions focused on the obvious limitations of sanitary measures, especially the stamping-out method. It was not only considered very expensive, and in some countries almost ruinous for the livestock sector, but also highly insufficient against the spread of a highly contagious disease such as foot-and-mouth disease, which easily crossed borders. Experts like Leclainche placed their hope in new efforts of immunisation, for which extensive research was in progress.[54]

The discussion about a more global perspective on the spread of diseases revealed further levels of complexity. Levinus de Blieck, a Dutch professor of veterinary medicine who had worked in the Dutch East Indies, emphasised the connections of European with Asian outbreaks of rinderpest as well as the differences between European and Asian emergences of the viruses. He argued that European veterinarians and administrations needed to give far more consideration to tropical diseases that were being introduced into Europe.[55] The Belgian veterinary inspector-general Henri de Roo pointed out that most European veterinarians (meaning those without experiences in the colonies) could no longer even recognise the symptoms of dangerous diseases that had become rare, such as rinderpest, which were now travelling the world and reappearing in Europe.[56] The debate also raised the issue of using European veterinary concepts and practices in non-European countries and in the colonies. Especially the method of stamping-out was contested in a global context when pathways of diseases were variable, or when territorialities were shifting. De Blieck pointed out to consider more closely local conditions with regard especially to tropical countries. Leclainche agreed; in some African countries, such as Egypt, he stated, the method had been given up because it had not shown much effectiveness.[57] But the method also created problems in Eastern Europe. The bacteriologist Julian Nowak, politician and rector of the Jagiellonian University in Krakow, who would in 1922 briefly become prime minister of Poland, outlined how Poland was constantly under threat of rinderpest outbreaks because of permeable borders with Bolshevik Russia.[58] Nowak argued that, with a large proportion of animals already destroyed because of the war and its consequences, obligatory culling not only of sick but also of animals suspected of rinderpest would have disastrous economic consequences for the population.[59] The measure of stamping-out, controversial but for decades considered one of the universal tools against epizootics, already displayed visible limitations in its practical use in Europe, especially when territorial borderlines were shifting or difficult to be controlled. Some conference participants still advocated rigorous culling as the preferable instrument; Hutyra phrased the measure for the discussion of a resolution on the fight against rinderpest: "In Europe, compulsory culling of sick or suspect cattle and also, as broadly as possible, of contaminated animals that appear healthy, with large and immediate compensation".[60]

It was the delegate from Haiti, the historian and diplomat Louis-Dantès Bellegarde, who took the opportunity to criticise the omnipresence of European viewpoints and ask for a more global perspective on the problem area. He had just been appointed as Haitian minister to Paris. Soon, he would become a key figure of the pan-African movement, praised in 1926 by W. E. B. Du Bois as the "international spokesman of black folk".[61] Dantès Bellegarde, who was not a veterinary expert, added a fresh perspective to the discussions. Diplomatically, he admitted that he did not understand most of the detailed debates. But he wanted to know why the relevant measures against epizootics such as rinderpest—especially the method of culling—seemed to be tailored to the conditions in Europe:

> I believe that the Conference considers culling to be the safest method to contain the spread of rinderpest; but I also wonder about the reason for this limitation [of the method of culling in the proposed proposal] to Europe. We are an international conference. Rinderpest can occur in countries outside Europe. If culling seems to be the safest way to control rinderpest, why limit it to Europe?[62]

The answers to his question about the focus on Europe, and what to agree upon for other countries with regard to the question of culling in the case of rinderpest, were mixed. Hutyra pointed out that the dangers of rinderpest for Europe were severe and therefore the issue was of high priority, and mandatory culling made no sense in countries where the disease was enzootic.[63] Leclainche added that in India, the method was impossible to be implemented on a broad scale for religious and cultural reasons.[64] Eventually, the conference agreed on a formulation of compromise in their final resolutions on rinderpest, which stated that obligatory culling was, "in principle" (replacing the phrase "in Europe"), one of the major practices to be of usage in case of sick, of suspect and of contaminated animals, and they included the recommendation of compensation.[65] But the problem remained that the global dimension of adequate responses against highly contagious epizootics such as rinderpest and foot-and-mouth disease was insufficiently addressed in the discussion. Vaccination methods against rinderpest and a number of other diseases existed, but they were expensive and did not always have the desired effect. One outcome of the discussion was that European concepts and practices had multiple limitations, which became particularly apparent when they were applied on a global scale, and that alternatives or conceptual enlargements were underdeveloped.

The experts were more in agreement on the question of setting up a centralised information transfer on animal diseases. The main problem was the lacking quality of many national animal disease bulletins, if they existed at all, and their poorly organised distribution. This situation did not much help to increase certainty about the respective disease situation. Information about new disease outbreaks tended to arrive too late, even in neighbouring countries. Among other things, the experts criticised the slowness of the diplomatic channels of communication, and advocated direct communication lines between the veterinary administrations.[66] To improve the national bulletins, the conference participants laid down a number of principles to make them useful for informing the veterinary services of other countries, in particular the list of diseases that should be included and the geographical dimension of disease outbreaks.[67] On the question of how the security of animal exports could be increased, the participating experts emphasised the need for certificates of origin, issued and controlled by state veterinary services in exporting countries. A later conference should determine a suitable formulation of the content.[68]

One of the main points on the agenda of the conference was the initiative for the new international office on the topic of epizootics. According to the French plans, the office was intended to provide through its work an increased level of mutual security in the matter of epizootics and contagious diseases of animals. This objective was only possible if as many states and colonies as possible cooperated in the project. The question was how to to set up, finance and manage the office and how to organise its international scientific-veterinary profile to ensure its relevance. The suggested working programme of the office followed in many aspects the debates at the pre-war international veterinary congresses.[69] The new office should above all organise a new systematic form of information transfer on animal diseases. In its international bulletin, it would inform regularly about the global situation of animal diseases, especially contagious ones, by providing statistics of disease cases, reports about the respective situation of veterinary control, and information on developments in research.[70] The organisation of a centralised information transfer on epizootics on a worldwide scale through a bulletin was dependent on sufficient communication infrastructures as well as on the close contacts with national veterinary administrations and research institutions. Therefore, an important point in the debate at the conference was the question of whether the new office should be independent, or be attached to an existing international organisation. Together with several other experts, Pottevin advocated an attachment to the OIHP.[71] As a longtime official of the OIHP, he had just co-headed an international enquiry of the OIHP on the pandemic of influenza and was well acquainted with the necessities of international work on contagious diseases.[72] The OIHP was an international body situated in Paris with a structure and thematic tasks that seemed in many aspects to be similar to those envisaged for the new office. The idea of attachment to the OIHP offered some structural advantages. A number of countries which were not involved in the League of Nations were members of the OIHP, in particular the United States. Before the war, Pottevin argued, the structure of the OIHP had been quite flexible for integrating experts and their agendas even if their home country was not a member.[73] But not everyone shared Pottevin's argumentation. In the view of Albert Calmette, the number of diplomats at the OIHP was too large in relation to the technical experts, the new veterinary office should instead be composed mainly of technical experts.[74] There was also a veterinary-scientific argument for attachment to the OIHP, which led to a discussion about the new office's scope of activities and responsibilities. Calmette pointed out that any new office working on animal diseases needed to be closely connected to institutions that were occupied with human disease prophylaxis in order to also include zoonotic diseases.[75] De Jong, professor of comparative pathology in Leiden, argued too for a stronger focus of the new office on zoonoses. The disciplinary separation of the two branches of human and animal medicine, he outlined, only created the problem of where common diseases of humans and animals were to be internationally addressed. De Jong outlined that the focus on diseases of agricultural relevance could therefore be only one of the new office's tasks:

> It is not possible to separate the health of humans from that of animals.... We have an international office on public hygiene regarding human diseases. One of the goals of veterinary science is also to prevent hazards to humans from animal diseases. You do not want to fight all infectious diseases, you only target infectious diseases that are economically dangerous; but there are diseases that are dangerous to human health, and that seems to me to be of great importance.[76]

However, the question of including explicitly a focus on zoonotic diseases and their consequences in international work was not followed for the moment.

The planned office was undoubtedly an institution oriented on European concepts, practices, concerns and requirements. The inducement for its establishment had been the threats to the European livestock economy; its tasks were based on European conceptions of veterinary services and priorities; and its member states were self-evidently expected to accept European standards of veterinary services and organise their systems of communication and information transfer along these lines. The question of whether Europe also exported diseases to other areas of the world, for example to the Americas or to its colonies, was not a major issue of discussion, although ships from Europe had, for instance, introduced foot-and-mouth disease at several occasions into Argentina and Uruguay in the late 19th century.[77] Moreover, the Portuguese representative, Paulo Nogueira, professor at the Lisbon veterinary school, pointed out that the conference needed to consider more the inter-European differences in veterinary approaches, as well as the close involvement of Europe with other parts of the world through the links with colonies: "Portugal is not, in the European civilisation, as advanced as, for example, France or Germany. We do not have a perfect veterinary organisation... a poor country cannot have a perfect organisation".[78] Nogueira's argumentation demonstrated the constructed limited meaning of 'Europe' for international veterinary debates—the term referred, above all, to the demands of Western and Central European countries, France, Britain, Belgium, the Netherlands, Denmark, Germany, Austria, and Italy, and far less to large parts of Southern and Southeastern Europe; the discussion also included the animal disease problems of several Eastern European countries with their borderlands to Russia, especially when they concerned Western Europe.

For most participants, the conference was a remarkable success; the delegates had in four days managed to draft the framework and working programme of a new international office. The existence of a similar international body in the shape of the OIHP had certainly accelerated the process of creating a new international office concerned with the containment of epizootics. As a response to the situation of epizootics and to prepare the new institution, the resolutions of the conference included a number of recommendations for the consideration of the governments with regard to future animal disease control and international cooperation.[79] Firstly, in the tradition of the pre-war international congresses, the resolutions proposed several recommendations for the improvement of the control of rinderpest, dourine and foot-and-mouth disease. In the case of rinderpest, they emphasised the continuing uncertainties regarding the species susceptible to the disease, which justified prohibition measures in the case that pigs and ruminants were imported from regions that could be affected by the disease. They further underlined the necessity of comprehensive research on the rinderpest virus, its nature and its spread. The resolutions included recommendations on the principles of rinderpest containment: the necessity of immediate telegraphic notification of disease cases to neighbour countries, the principle of obligatory culling and recompensation of livestock owners, the prohibition of using virulent material for immunisation efforts, and the production of vaccines only in state-controlled scientific institutions.[80] In the case of foot-and-mouth disease, they also recommended building up close scientific relations and exchange between research institutions working on immunisation methods.[81] For the containment of dourine, the resolutions outlined the necessity of further research activities or research, and in-depth surveillance of the disease situation by the veterinary services.[82] Secondly, the resolutions included recommendations concerning a regular sanitary information transfer ("renseignements sanitaires et bulletins sanitaires").[83] Again similar to the very first international veterinary congresses in the 1860s, the conference established a list of diseases to be used for information exchange via national bulletins, as well as instructions on the statistical survey. The list included economically relevant diseases and several zoonoses—rinderpest, foot-and-mouth disease, bovine pleuropneumonia, anthrax, sheep pox, rabies, glanders, dourine and swine fever.[84] Thirdly, the resolutions stated that exporting countries needed to establish unified sanitary certificates for traded animals that were issued by a professional state veterinary service; a later meeting would discuss the details.[85] Fourthly and most importantly, the resolutions recommended to establish a new international office with its seat in Paris. Its three main tasks were described as, first, the regular collection of information on and documentation of the animal disease situation in as many countries as possible, and the distribution of information to the participating governments; second, providing input and coordination for new research regarding thematic questions on which international collaboration would be of essence; and third, preparing projects of international conventions to advance intergovernmental cooperation in the field.[86] The resolutions explicitly mentioned that the office should have an exclusively scientific character. Its international committee and its directorate would consist of experts ("délégués techniques")[87] sent by member states, not politicians or diplomats. Subject to agreement by the participating governments, it should be attached to the OIHP.[88]

In total, the international conference of 1921 was an immense step toward establishing new international efforts to control animal diseases on a more permanent basis than before. The support of governments finally made a new institution of veterinary internationalism possible, which also meant more recognition for the veterinary agenda and new financial support from the governments. The measures discussed at the conference aimed to establish higher levels of mutual security concerning animal diseases through the repertoire of internationalism, channeled and centralised by the new office—a new system of information transfer, scientific exchange through established international meetings and administrative-scientific networks, and close international collaboration in veterinary research of diseases and prevention methods. While the conference was definitely a fresh start to veterinary internationalism, most of its core ideas and points of discussion had already been debated at the international veterinary congresses before the war. The awareness of the role of global connections for the spread of diseases did not bring about a thorough change of the perspective and of methods. Veterinary problems in non-European states and colonies were not central issues in the discussions. The integration of non-European participants in the conference underlined its international character and, in practice, helped to increase the net of veterinary contacts and thereby the amount and reliability of global disease information. There was, however, a difference to 19th-century veterinary internationalism in the more active debate on the tensions between the global problem area and the European orientation of veterinary concepts. The discussions on the tasks and thematic focus of the new international office, the future OIE, showcased the ongoing problems of Europe-focused conceptual and institutional orientations. A more global approach in terms of institutional structures, concepts, and practices, for instance also considering enzootic diseases and zoonoses, was not the main focus of the meeting.

The creation of the Office International des Epizooties (OIE), 1921–1927

After the international conference of 1921, the Ministry of Foreign Affairs prepared the draft text of an international convention together with statutes of the new office. The process of organising the international office started with some confusion caused by a similar international project. Shortly after the conference of May 1921, the French Ministry of Foreign Affairs was informed that a new international office for the study and containment of foot-and-mouth disease had been founded in Buenos Aires.[89] The correspondence demonstrated again that there existed considerable gaps in diplomatic and professional communication about the issue of animal diseases.

It also showed the still unclear position and fluctuating competencies of experts in the context of international events. It turned out that in September 1920, an international congress on foot-and-mouth disease (Congreso Internacional de Fiebre Aftosa) had taken place in Buenos Aires. It was attended by representatives from Germany, Argentina, Brazil, Cuba, Chili, El Salvador, Spain, France, Paraguay, Peru, Switzerland, and Uruguay.[90] The congress founded the International Institute of Foot-and-Mouth Disease (Instituto Internacional de la Fiebre Aftosa) in Buenos Aires. It was intended to be a new international institute with its structure based on the model of the IIA and the International Institute of Refrigeration (IIR). Funded by the member states, its areas of responsibility appeared strikingly similar to those of the envisaged Paris office: the centralised documentation and distribution of scientific information, worldwide statistical coverage of disease cases, regular notifications in case of outbreaks, and the implementation of a worldwide telegraphic alarm service. The project, concentrating on one particular disease, planned to go even further than that of the planned office in Paris—the institute envisaged the establishment of an international laboratory for the research, production and worldwide distribution of a new foot-and-mouth-disease vaccine.[91] This activity would, if successful, have created a new and path-breaking effort to address an animal disease through centrally coordinated international activities between a number of South American and European countries. A transboundary programme of vaccination—should a suitable vaccine have been produced—would also have been linked to a new concept of joint international animal disease control beyond the territorial nation state framework. The French Minister of Agriculture, Lefebvre du Preÿ, wondered why nobody had mentioned the project at the Paris conference of May 1921. He rejected French participation in the Buenos Aires project because of its interference with the plans of the new international office in Paris.[92] Further, he worried that the French expert who had attended the Buenos Aires conference, the French-Argentinian veterinarian José Lignières, director of the Argentinian National Institute of Bacteriology, had assumed some form of official responsibilities and signalised French official participation in the project.[93] Correspondence between several French ministries, the consulate in Buenos Aires and Lignières himself showed that there was confusion about the role and competencies of experts in international diplomacy. Veterinary experts such as Lignières were present as national representatives on international platforms, and their status called into question the boundaries between official assignment, semi-official technical advisory position, and professional but non-official participation as a member of a scientific community. The Buenos Aires congress, following common practice, listed Lignières together with other participants as national delegate. To clear up the confusion, he argued that he had acted without an official mandate.[94] The new international institute in Buenos Aires was indeed established in 1920. It conducted a series of collaborative laboratory experiments in the following years concerning foot-and-mouth disease immunisation and published the results, but did not attract much international attention or appear in other international contexts during the following years.[95]

The draft of the international convention for the new international office in Paris, accompanied by statutes, was based on the resolutions of the conference of 1921 and arranged the aims, principles of organisation and the legal modalities of participation. The office was intended to be an intergovernmental bureau with its seat in Paris. As was usual for international bureaux, it was officially independent from French authority. Article 4 of the statutes included a description of the objectives of the new office referring to the invitation circular of the conference of 1921. It pointed to the opportunity to initiate new research activities and increasing international collaboration. It was formulated in an open way, allowing a certain flexibility for activities beyond the tasks of organising the information transfer.[96] Article 6 emphasised that the governing committee of the new office had to be composed of technical experts appointed by the governments, which underlined the technical role and supposedly non-political character of the new office.[97] The envisaged monthly international bulletin of the office, as described in Article 10, should encompass notifications and legal arrangements, statistics of disease cases, information on research news and on relevant veterinary publications.[98] The relations of the new Paris office to other international bodies were indeed a central issue, not least because the problem area of animal diseases concerned numerous other national and international actors. Article 17 of the draft statutes mentioned a direct attachment of the new office to the OIHP, pending approval by the OIHP member governments.[99] There was still also the option to attach it to the League of Nations, to complement its work in the field of public health, and benefit from its networks of official and non-official international activities as well as from its funds. The League was beginning to address subject areas that were related to those of new office of epizootics—epidemiological intelligence, projects of standardisation, organising and documenting research on specific disease problems, prevention, and projects of public health. Especially the British Foreign Office insisted on clarification on the point of an attachment to the League, referring to Article 24 of the League Covenant.[100] However, the affiliation of the new office to an already existing international institution would have closed some important doors in terms of memberships. Most significantly in case of an attachment, non-member states of the League or the OIHP would not officially participate in the new office. The French Ministry of Foreign Affairs insisted that the member governments needed to decide on the independence of the institution or its attachment to another international organisation.[101] In spring 1922, the draft texts of the convention and the statutes of the OIE were communicated to the governments that had sent a delegation to the conference of 1921. It soon became clear that there was no consensus among the member governments of the OIHP for the attachment of a new institute on animal diseases. In November 1922, the French Ministry of Foreign Affairs suggested removing references to the attachment to the OIHP from the statutes and convention, fearing that the whole convention would fail.[102] Though in the ensuing diplomatic correspondence in the first half of 1923 many governmental officials expressed regret about this decision, the modification was accepted.[103]

The organisers of the 1921 conference had underlined that the new international project for animal disease control required the participation of as many member states as possible. Diplomatic correspondence from the French Ministry of Foreign Affairs during the process of setting up the new office again highlighted the argument of many governments not being able to participate for financial reasons or because of the lack of the required standards of national veterinary services. The significance of the project was commonly accepted, at least in the correspondence, but especially governments of smaller states were hesitant to spend money on a project that seemed to be a long-term investment for the future. The French organisers, again taking the example of the OIHP, established a quota system of six categories of payments for member states. The governments could choose their category and therefore the amount of financial contribution.[104] Despite this flexible system of payment categories, allowing also smaller and financially constrained countries to participate, the intended broad list of members in the new office, both within and beyond Europe, was not easy to be realised. The Austrian government explained that, for financial reasons, Austria could presently not afford to participate, even though the country had an important livestock sector and a significant interest in the project.[105] The government of Haiti regretted to not be able to participate at the present moment.[106] Other governments, such as that of Ecuador, pointed out that they did not provide state-organised veterinary services, and could therefore not send the requested regular information.[107] Colonial or mandate governments as well as other dependent territories joined independently.[108] A serious problem for the new office was the non-participation of countries with vast territories that were significant for the spread of epizootics among domestic animals and wildlife, as well as those which provided an important livestock husbandry sector. China, Australia, the United States and Canada did not join. The U.S. government argued that the respective diseases in the focus of the new office did not appear on its territory, and that the United States would not have any advantage from signing the convention for a new international office.[109] Russia, since late 1922 the Soviet Union, had not been represented at the 1921 conference. Given the significance of the disease situation of Russia, especially for the countries of Eastern Europe and the vast devastations during the civil wars, participation would be important. In late 1924, the Soviet government began to show interest in the project of the international office. France had taken up diplomatic relations with the Soviet Union in the meantime, and the Soviet embassy in Paris asked for information about the project and its legal background.[110]

The international agreement for the new Office International des Epizooties (OIE) was established for signing on January 25, 1924. Twenty-eight governments from Europe, Latin America, Asia and Africa signed the convention—Argentina, Belgium, Brazil, Bulgaria, Czechoslovakia, Denmark, Egypt, Spain, Finland, France, Great Britain, Greece, Guatemala, Hungary, Italy, Luxemburg, Morocco, Mexico, the Principality of Monaco, the Netherlands, Peru, Poland, Portugal, Romania, Siam, Sweden, Switzerland and Tunisia; in May 1924 those of Ireland and British India also signed.[111] In 1927, the office took up its provisional residence in the building of the French Ministry of Agriculture, and in March 1927, the International Committee of the new OIE held its first constituting meeting.[112] Of interest were the categories of annual payments which the new members chose. As a response of the French Ministry of Foreign Affairs to an enquiry of the Embassy of Japan outlined in 1929, the Soviet government paid the highest possible financial contribution; so too did, as the diplomatic correspondence explained, only Germany, Britain, France and Italy.[113] A higher contribution did not formally lead to more influence on the working proceedings and decisions of the OIE, it was a question of prestige, and probably governments thought about an option of future governmental engagement in the OIE. With the instalment of the OIE, a first institutional approach toward addressing animal disease problems beyond national borders was realised, based on the late–19th-century model of an international office. In contrast to the OIHP, however, the meetings of its international committee should be dominated by veterinary experts. The new institution, though in structural terms rooted in the 19th century, reflected the transformation of internationalism and diplomacy that had already started before the war. The creation of the new office underlined the ideal of advancing technical and scientific solutions for urgent problem areas in an international atmosphere that was constructed to be detached from political conflicts—even though the subject of animal diseases was a politically and economically highly important thematic area. The position of veterinary experts had gained more significance in international relations. They had managed to underscore their interpretation of international cooperation by insisting on their professional authority relevant for transboundary problems, defined to be of a technical-scientific nature. With their new institutional platform, they clearly hoped to advance new options in international animal disease control. The project of the new office offered the chance to establish a permanent global network of information transfer and scientific exchange, and, even more so, it also made it possible to promote research projects on animal diseases. This was a promising step towards a new, systematic collaboration to deal with a global problem area, especially in times of continuing political and economic tensions such as the 1920s and 1930s. However, the international conference of 1921 and the founding process of the OIE already revealed several future problems of international activities in the field of animal diseases. The discussions during the organisation of the new office showed that engagement in international work was a tool for governments that could afford the financial resources and infrastructures for animal disease control. This was a serious problem for the future scope of international veterinary work, and also for the OIE, which depended on international engagement, the funding of member states, and on a worldwide network of cooperation to maintain an effective system of disease information. Furthermore, the discussions reflected again the question of integration and exclusion in international projects concerning globally relevant problems. The spread of rinderpest—with an immediate danger for Europe—and of foot-and-mouth disease in Europe were the main motivation for the conference of 1921, and the organisers of the new office tried to realise a global scope of member states through their system of different financial contributions. Most discussions of 1921 and the following years, however, still centered on European problems and strategies. Questions about the relationship between animal diseases and public health, and the globally dangerous potential of zoonotic diseases, were included to some extent, but were not a main focus of the new international office, nor of the OIHP or the IIA. Other international organisations, such as the League of Nations and the ILO, would approach some of these topics during the following years.

Notes

1 Madeleine Herren, *Internationale Organisationen seit 1865. Eine Globalgeschichte der internationalen Ordnung* (Darmstadt: Wissenschaftliche Buchgesellschaft, 2009), 55–8.

2 M. Dorset, C. N. McBryde, and W. B. Niles, 'Remarks on "Hog" Flu', *Journal of the American Veterinary Medical Association* 62 (1922): 162–71; on the discussion see: Stacey Schultz-Cherry, Christopher W. Olsen, and Bernard C. Easterday, 'History of Swine Influenza', in *Swine Influenza*, ed. Jürgen A. Richt and Richard J. Webby (Berlin, Heidelberg: Springer, 2013), 22–3. About the influenza pandemic of 1918–20 in historiography see: Tamara Giles-Vernick, Susan Craddock, and Jennifer Lee Gunn, eds., *Influenza and Public Health: Learning from Past Pandemics* (London, Washington DC: Earthscan, 2010); Howard Philips, 'The Re-Appearing Shadow of 1918: Trends in the Historiography of the 1918–19 Influenza Pandemic', *Canadian Bulletin of Medical History = Bulletin Canadien d'Histoire de la Médecine* 21, no. 1 (2004): 121–34.

3 Robert J. Stordy, 'Belgium and East Africa. Danger to a Great Cattle Area. Letters to the Editor', *The Times*, May 24, 1919, 8. The Scotsman Robert J. Stordy (1873–1943) worked since 1898 in the British East African Protectorate; see: Irina Podgorny, 'A Horse-Cloth for Uganda, or How an Account by a Transhumant Veterinary Connects Histories, Animal Diseases and Continents', *History and Technology* 34, no. 1 (2018): 71–8.

4 Stordy, 'Belgium and East Africa', 8.

5 'The Risk of Rinderpest. Belgian Measures in East Africa', *The Times*, May 28, 1919, 11. The following day, *The Times* published a commentary which strongly supported Stordy's argumentation. See: 'A Word to Belgium', *The Times*, May 29, 1919, 13.

6 Joseph-Honoré Ricard (1880–1948) was French Minister of Agriculture in 1920–1921 and author of several books on topics of agricultural financing and insurance. For Ricard's biography see the page of the French Ministry of Agriculture: 'Agriculture.gouv.fr: Les Ministres de l'Agriculture – IIIe République (1870–1940). Joseph-Honoré Ricard', accessed March 19, 2020, http://agriculture.gouv.fr/220-Joseph-Honore-RICARD.

7 Joseph Ricard, *Les Origines de la Première Conférence Internationale des Epizooties, Fièvre Aphteuse, Peste Bovine* (Paris: de Chaix, 1921), 4–5.
8 Ricard, *Origines de la Première Conférence Internationale des Epizooties*, 4–5.
9 Ricard, *Origines de la Première Conférence Internationale des Epizooties*, 4–6.
10 Ricard, *Origines de la Première Conférence Internationale des Epizooties*, 7.
11 Ricard, *Origines de la Première Conférence Internationale des Epizooties*, 7.
12 Ricard, *Origines de la Première Conférence Internationale des Epizooties*, 7.
13 Ricard, *Origines de la Première Conférence Internationale des Epizooties*, 7.
14 Ricard, *Origines de la Première Conférence Internationale des Epizooties*, 7–8 (translation: CK).
15 William P. Taylor, Peter L. Roeder, and Mark M. Rweyemamu, 'History of Vaccines and Vaccination', in *Rinderpest and Peste des Petits Ruminants. Virus plagues of large and small ruminants*, ed. William P. Taylor, Thomas Barrett, and Paul-Pierre Pastoret (London and Burlington, MA: Elsevier/Academic, 2006), 223–5.
16 Ricard, *Origines de la Première Conférence Internationale des Epizooties*, 8. About the Serum Institute in Abbassia (Cairo): Charles Todd, *Description of the Serum Institute (Abbassia) and of the Preparation of Cattle Plague Serum* (Cairo: Govt. Press, 1912).
17 Ricard, *Origines de la Première Conférence Internationale des Epizooties*, 9–10.
18 About the measures in Belgium and the French-Belgian cooperation: Marc Mammerickx, 'La Peste Bovine, Jules Bordet et le Centre Sérumigène de Cureghem', *Annales de Médecine Vétérinaire* 147 (2003): 200–3. The production of the serum by the French-Belgian team was described in detail in a publication of the Alfort researchers Eugène Nicolas and Paul Rinjard: Eugène Nicolas and Paul Rinjard, *La Production du Sérum contre la Peste Bovine au Centre Sérumigène de Cureghem-lez-Bruxelles* (Toulouse: Imprimerie Ouvrière, 1922).
19 Ricard, *Origines de la Première Conférence Internationale des Epizooties*, 8–9.
20 Ricard, *Origines de la Première Conférence Internationale des Epizooties*, 10.
21 Ricard, *Origines de la Première Conférence Internationale des Epizooties*, 11 (translation: CK).
22 Ricard, *Origines de la Première Conférence Internationale des Epizooties*, 11.
23 Ricard, *Origines de la Première Conférence Internationale des Epizooties*, 10–11.
24 For an overview on the League of Nations' early work on public health see the material from its Information Section: League of Nations Information Section, *The Health Organisation of the League of Nations* (Geneva: League of Nations, 1923). For a detailed historical analysis and interpretation: Iris Borowy, *Coming to Terms with World Health. The League of Nations Health Organisation, 1921–1946* (Frankfurt am Main and New York: Peter Lang, 2009).
25 For biographical information about the career and research of Emmanuel Leclainche (1861–1953) see: Laure Alnot and Guilhem Pistre, 'La Vie et l'Oeuvre d'Emmanuel Leclainche (1861–1953)', *Bulletin de la Société Française d'Histoire de la Médecine et des Sciences Vétérinaires* 4, no. 1 (2005): 94–103.
26 Alnot and Pistre, 'Emmanuel Leclainche', 98 (translation: CK).
27 Ministère des Affaires Etrangères (France), Le Président du Conseil, Ministre des Affaires Etrangères, à Messieurs les Agents Diplomatiques de France, signed Leygues, November 12, 1920, 1–2 (translation: CK). ADLC: Unions, 429QO/332 (Office International des Epizooties, 1920–23). On the circular also: Cornelia Knab, 'Infectious Rats and Dangerous Cows: Transnational Perspectives on Animal Diseases in the First Half of the Twentieth Century', *Contemporary European History* 20, no. 3 (2011): 293.

28 Ministère des Affaires Etrangères (France), Le Président du Conseil, Ministre des Affaires Etrangères, à Messieurs les Agents Diplomatiques de France, signed Leygues, November 12, 1920, 3 (translation: CK). ADLC: Unions, 429QO/332 (Office International des Epizooties, 1920–23). The programme is also printed in: République Française, Ministère de l'Agriculture, ed., *Conférence Internationale pour l'Etude des Epizooties. Paris (25–28 Mai 1921). Compte Rendu Sténographique* (Paris: Imprimerie Nationale, 1921), 3–4.

29 Ministère des Affaires Etrangères (France), Le Président du Conseil, Ministre des Affaires Etrangères, à Monsieur le Ministre de l'Agriculture, February 4, 1921. ADLC: Unions, 429QO/332 (Office International des Epizooties, 1920–23).

30 Ministère des Affaires Etrangères (France), Le Général Pellé, Haut-Commissaire de la République Française en Orient, à Son Excellence Monsieur Aristide Briand, le Président du Conseil, Ministre des Affaires Etrangères, April 8, 1921; Le Président du Conseil, Ministre des Affaires Etrangères, à Monsieur le Haut-Commissaire de la République à Constantinople, April 24, 1921. ADLC: Unions, 429QO/332 (Office International des Epizooties, 1920–23).

31 Ministère des Affaires Etrangères (France), Direction des Affaires Politiques et Commerciales, Note pour la Direction des Affaires Administratives et Techniques, signed Emmanuel de Peretti de la Rocca, April 4, 1921. ADLC: Unions, 429QO/332 (Office International des Epizooties, 1920–23).

32 Ministère des Affaires Etrangères (France), Le Ministre des Affaires Etrangères à Ambassadeur Français [sic], Rome, Londres, Bruxelles, signed Briand/Maurice Herbette, April 11, 1921. ADLC: Unions, 429QO/332 (Office International des Epizooties, 1920–23); Deutsche Botschaft in Frankreich (translation) to the Ministère des Affaires Etrangères (France), May 22, 1921. ADLC: Unions, 429QO/332 (Office International des Epizooties, 1920–23).

33 Foreign Office (United Kingdom), George Villiers to His Excellency le Comte de Saint-Aulaire (Ambassador of France, London), February 22, 1921. ADLC: Unions, 429QO/332 (Office International des Epizooties, 1920–23).

34 Ministère de l'Agriculture (France), Le Ministre de l'Agriculture à Monsieur le Président du Conseil, Ministre des Affaires Etrangères, March 23, 1921. ADLC: Unions, 429QO/332 (Office International des Epizooties, 1920–23).

35 Légation de France au Centre Amérique, Le Chargé des Affaires de la République Française au Centre Amérique au Ministre des Affaires Etrangères, February 10, 1921, with translation of a letter of the Foreign Ministry of Guatemala. ADLC: Unions, 429QO/332 (Office International des Epizooties, 1920–23).

36 Ministère des Affaires Etrangères (France), Direction des Affaires Administratives et Techniques, Note pour la Direction des Affaires Politiques et Commerciales (Maurice Herbette à Emmanuel de Peretti de la Rocca), May 25, 1921. ADLC: Unions, 429QO/332 (Office International des Epizooties, 1920–23).

37 United Nations Treaty Collection, 'Pacte de la Société des Nations/Covenant of the League of Nations', United Nations Treaty Collection: League of Nations Treaty Series, accessed March 1, 2021, https://treaties.un.org/doc/source/covenant.pdf (Article 24).

38 Ministère des Affaires Etrangères (France), Direction des Affaires Administratives et Techniques, Note pour la Direction des Affaires Politiques et Commerciales (Maurice Herbette à Emmanuel de Peretti de la Rocca), May 25, 1921, 2–3. ADLC: Unions, 429QO/332 (Office International des Epizooties, 1920–23).

39 On the conference see also: Amanda Kay McVety, *The Rinderpest Campaigns: A Virus, Its Vaccines, and Global Development in the Twentieth Century* (Cambridge: Cambridge University Press, 2018), 13-15, 33; Knab, 'Infectious Rats and Dangerous Cows', 293–5; Mark Harrison, *Contagion: How Commerce Has Spread Disease* (New Haven: Yale University Press, 2012), 226–7.

40 The Pasteurian Émile Roux (1853–1933), renowned for his works on diphtheria therapy, had also, together with Edmond Nocard, described for the first time the mycoplasma pathogen of bovine pleuropneumonia in 1898. For a short biography see: Institut Pasteur, 'Institut Pasteur – Archives: Dossiers Biographiques. Emile Roux (1853–1933)', accessed June 1, 2020, https://webext.pasteur.fr/archives/rou0.html. Albert Calmette (1863–1933) is today especially known for his development of the tuberculosis vaccination (BCG) together with Camille Guérin. For a short biography see: Institut Pasteur, 'Institut Pasteur – Archives: Dossiers Biographiques. Albert Calmette (1863–1933)', accessed April 1, 2021, https://webext.pasteur.fr/archives/cal0.html. About the physician and biologist Henri Pottevin (1865–1928), who worked in the Institut Pasteur and the OIHP (director 1924–1928) see the biographical information of the Pasteur Institute: Institut Pasteur, 'Institut Pasteur: Repères Chronologiques. Henri Pottevin (1865–1928)', accessed April 1, 2021, https://webext.pasteur.fr/archives/pot0.html.

41 The veterinarian and microbiologist Henri Vallée (1874–1947) worked in the 1920s in Alfort on the development of foot-and-mouth disease vaccination. See: Jean Blancou and Claude Meurier, 'Brève Histoire de Quelques Laboratoires Nationaux Vétérinaires Français: Première Partie', *Bulletin de la Société Française d'Histoire de la Médecine et des Sciences Vétérinaires* 6 (2006): 79–107.

42 The list of represented states, dominions and colonies included: Germany, USA, Argentina, Austria, Belgium, Bolivia, Brazil, Bulgaria, Chile, Denmark, Dominican Republic, Ecuador, Spain, Finland, Great Britain, Ireland, Canada, South Africa, Australia, India, New Zealand, Greece, Haiti, Hungary, Italy, Japan, Luxemburg, Morocco, Monaco, Nicaragua, Norway, Paraguay, Netherlands, Peru, Poland, Portugal, Romania, the Kingdom of Serbs, Croates and Slovenes, Sweden, Switzerland, Czechoslovakia, Tunisia and Venezuela. République Française, Ministère de l'Agriculture, *Conférence Internationale pour l'Etude des Epizooties 1921*, Liste des Etats, 4; Delegués, 6–10.

43 République Française, Ministère de l'Agriculture, *Conférence Internationale pour l'Etude des Epizooties 1921*, 5–6, 10–12.

44 Hutyra, in: République Française, Ministère de l'Agriculture, *Conférence Internationale pour l'Etude des Epizooties 1921*, 12–13. Hutyra mistakenly mentioned 1871 as the date of the Vienna rinderpest conference.

45 République Française, Ministère de l'Agriculture, *Conférence Internationale pour l'Etude des Epizooties 1921*, 10-63.

46 Leclainche, in: République Française, Ministère de l'Agriculture, *Conférence Internationale pour l'Etude des Epizooties 1921*, 14–15.

47 Leclainche, in: République Française, Ministère de l'Agriculture, *Conférence Internationale pour l'Etude des Epizooties 1921*, 15, 22–3.

48 Leclainche, in: République Française, Ministère de l'Agriculture, *Conférence Internationale pour l'Etude des Epizooties 1921*, 14–15.

49 At the Vienna conference on rinderpest in January 1921, the participating states were Czechoslovakia, Germany, Austria, Hungary, Poland, Romania, and the Kingdom of Serbs, Croats and Slovenes. Protocol Drawn up by the Conference of Central European States on the Eradication of Cattle-Plague, held at Vienna in January 1921 (January 17–18, 1921), 1 (translation). LON R2780, 10D/1477/309. See also: Hutyra, in: République Française, Ministère de l'Agriculture, *Conférence Internationale pour l'Etude des Epizooties 1921*, 19.

50 Protocol Drawn up by the Conference of Central European States on the Eradication of Cattle-Plague, held at Vienna in January 1921, 1–3 (translation). LON R2780, 10D/1477/309.
51 About the conference debates on rinderpest see also: McVety, *The Rinderpest Campaigns*, 33.
52 Leclainche, in: République Française, Ministère de l'Agriculture, *Conférence Internationale pour l'Etude des Epizooties 1921*, 16.
53 For instance: De Jong, de Blieck, and Vallée, in: République Française, Ministère de l'Agriculture, *Conférence Internationale pour l'Etude des Epizooties 1921*, 26–8.
54 Leclainche, in: République Française, Ministère de l'Agriculture, *Conférence Internationale pour l'Etude des Epizooties 1921*, 30–1; general discussion on foot-and-mouth disease: 30-39.
55 De Blieck, in: République Française, Ministère de l'Agriculture, *Conférence Internationale pour l'Etude des Epizooties 1921*, 18, 27. For biographical information about the Dutch veterinarian L. de Blieck (1878–1965) see the database of professors of Utrecht University: Universiteit Utrecht, 'Catalogus Professorum: Professoren van de Universiteit Utrecht sinds 1636: Prof. Dr. L. de Blieck', accessed March 19, 2020, http://profs.library.uu.nl/index.php/profrec/getprofdata/192/1/9/0.
56 De Roo, in: République Française, Ministère de l'Agriculture, *Conférence Internationale pour l'Etude des Epizooties 1921*, 27. About Henri de Roo (1861–1930), Chief veterinary officer in Belgium and in 1927 elected first president of the committee of the OIE, see: Mammerickx, 'La Peste Bovine, Jules Bordet et le Centre Sérumigène de Cureghem', 203–4.
57 De Blieck and Leclainche, in: République Française, Ministère de l'Agriculture, *Conférence Internationale pour l'Etude des Epizooties 1921*, 51.
58 Nowak, in: République Française, Ministère de l'Agriculture, *Conférence Internationale pour l'Etude des Epizooties 1921*, 19–21. About Julian Nowak (1865–1946) see: J. Buszko, 'Julian Nowak', in *Österreichisches Biographisches Lexikon 1815–1950* (Wien: Verlag der Österreichischen Akademie der Wissenschaften, 1978).
59 Nowak, in: République Française, Ministère de l'Agriculture, *Conférence Internationale pour l'Etude des Epizooties 1921*, 95.
60 République Française, Ministère de l'Agriculture, *Conférence Internationale pour l'Etude des Epizooties 1921*, 94 (translation: CK).
61 [W. E. B. Du Bois], 'Dantes Bellegarde: International Spokesman of Black Folk', *The Crisis* 31 (April 1926): 295–6. For Dantès Bellegarde's international activities in context of the League of Nations see: Glenda Sluga, *Internationalism in the Age of Nationalism* (Philadelphia: University of Pennsylvania Press, 2013), 66–9.
62 Dantès Bellegarde, in: République Française, Ministère de l'Agriculture, *Conférence Internationale pour l'Etude des Epizooties 1921*, 97 (translation: CK).
63 Hutyra, in: République Française, Ministère de l'Agriculture, *Conférence Internationale pour l'Etude des Epizooties 1921*, 97–8.
64 Leclainche, in: République Française, Ministère de l'Agriculture, *Conférence Internationale pour l'Etude des Epizooties 1921*, 98.
65 République Française, Ministère de l'Agriculture, *Conférence Internationale pour l'Etude des Epizooties 1921*, 117–21, 125–31, 134 (resolution).
66 République Française, Ministère de l'Agriculture, *Conférence Internationale pour l'Etude des Epizooties 1921*, 59–63.
67 République Française, Ministère de l'Agriculture, *Conférence Internationale pour l'Etude des Epizooties 1921*, 62–3.
68 République Française, Ministère de l'Agriculture, *Conférence Internationale pour l'Etude des Epizooties 1921*, 65–78.
69 Aims and working programme of the office: République Française, Ministère de l'Agriculture, *Conférence Internationale pour l'Etude des Epizooties 1921*, 109–12.

70 Discussion about the office: République Française, Ministère de l'Agriculture, *Conférence Internationale pour l'Etude des Epizooties 1921*, 79–92, 109–15.
71 Pottevin, in: République Française, Ministère de l'Agriculture, *Conférence Internationale pour l'Etude des Epizooties 1921*, 84.
72 Henri Pottevin, 'Rapport sur la Pandémie Grippale de 1918–1919 présenté au Comité Permanent de l'Office International d'Hygiène Publique', *Bulletin de l'Office International d'Hygiène Publique* 13 (1921): 125–81. About the OIHP studies on influenza: David M. Morens et al., 'An Historical Antecedent of Modern Guidelines for Community Pandemic Influenza Mitigation', *Public Health Reports* 124 (February 2009): 22–4.
73 Pottevin, in: République Française, Ministère de l'Agriculture, *Conférence Internationale pour l'Etude des Epizooties 1921*, 84.
74 Calmette, in: République Française, Ministère de l'Agriculture, *Conférence Internationale pour l'Etude des Epizooties 1921*, 88.
75 Calmette, in: République Française, Ministère de l'Agriculture, *Conférence Internationale pour l'Etude des Epizooties 1921*, 80.
76 De Jong, in: République Française, Ministère de l'Agriculture, *Conférence Internationale pour l'Etude des Epizooties 1921*, 87, 90–1 (translation: CK). See also: Knab, 'Infectious Rats and Dangerous Cows', 294.
77 Manuel A. Machado, *Aftosa: A Historical Survey of Foot-and-Mouth Disease and Inter-American Relations* (Albany: State University of New York Press, 1969), especially Chapter 1.
78 Nogueira, in: République Française, Ministère de l'Agriculture, *Conférence Internationale pour l'Etude des Epizooties 1921*, 85 (translation: CK).
79 République Française, Ministère de l'Agriculture, *Conférence Internationale pour l'Etude des Epizooties 1921*, 134–6.
80 République Française, Ministère de l'Agriculture, *Conférence Internationale pour l'Etude des Epizooties 1921*, 134.
81 République Française, Ministère de l'Agriculture, *Conférence Internationale pour l'Etude des Epizooties 1921*, 134.
82 République Française, Ministère de l'Agriculture, *Conférence Internationale pour l'Etude des Epizooties 1921*, 135.
83 République Française, Ministère de l'Agriculture, *Conférence Internationale pour l'Etude des Epizooties 1921*, 135.
84 République Française, Ministère de l'Agriculture, *Conférence Internationale pour l'Etude des Epizooties 1921*, 135.
85 République Française, Ministère de l'Agriculture, *Conférence Internationale pour l'Etude des Epizooties 1921*, 135–6.
86 République Française, Ministère de l'Agriculture, *Conférence Internationale pour l'Etude des Epizooties 1921*, 136.
87 République Française, Ministère de l'Agriculture, *Conférence Internationale pour l'Etude des Epizooties 1921*, 136.
88 République Française, Ministère de l'Agriculture, *Conférence Internationale pour l'Etude des Epizooties 1921*, 136.
89 Légation de la République Française en Argentine, Monsieur Clausse, Ministre de France en Argentine, à Son Excellence Monsieur le Ministre des Affaires Etrangères à Paris, May 31, 1921. ADLC: Unions, 429QO/498 (Etude de la Fièvre Aphteuse).
90 Légation de la République Française en Argentine, Monsieur Clausse, Ministre de France en Argentine, à Son Excellence Monsieur le Ministre des Affaires Etrangères à Paris, May 31, 1921. ADLC: Unions, 429QO/498 (Étude de la Fièvre Aphteuse). About the Congreso Internacional de Fiebre Aftosa: Nelson Magallanes, 'Ciento Veinte Años de Aftosa en Uruguay', *Veterinaria* 29, no. 120 (Dicembre 1993): 6–7.

91 Ministère de l'Agriculture (France), Le Ministre de l'Agriculture à Monsieur le Président du Conseil, Ministre des Affaires Etrangères, September 17, 1921, 1–2. ADLC: Unions, 429QO/498 (Etude de la Fièvre Aphteuse).

92 Ministère de l'Agriculture (France), Le Ministre de l'Agriculture à Monsieur le Président du Conseil, Ministre des Affaires Etrangères, September 17, 1921, 3. ADLC: Unions, 429QO/498 (Etude de la Fièvre Aphteuse).

93 [Ministère des Affaires Etrangères (France)], Le Président du Conseil, Ministre des Affaires Etrangères, à Monsieur le Ministre de France en Argentine, March 15, 1922. ADLC: Unions, 429QO/498 (Etude de la Fièvre Aphteuse). About the veterinarian José Lignières (1868–1933): Academia Nacional de Agronomia y Veterinaria, 'Lignières, José', accessed March 19, 2021, http://www.anav.org.ar/index.php?option=com_content&view=article&id=190:lignieres-jose-med-vet&catid=43:de-numero-fallecidos&Itemid=69.

94 José Lignières to Roger Clausse, Ministre de France à Buenos Aires, April 18, 1922. ADLC: Unions, 429QO/498 (Etude de la Fièvre Aphteuse).

95 About the Instituto Internacional de la Fiebre Aftosa in Buenos Aires: Pedro Bergés, *Instituto Internacional de la Fiebre Aftosa y Oficina Sanitaria Internacional. Trabajos realizados para su Creación en Buenos Aires (1920–1924)* (Buenos Aires: Revista de Industria Lechera y Ganaderia, 1924); Magallanes, 'Ciento Veinte Años de Aftosa en Uruguay', 7.

96 République Française, Ministère de l'Agriculture, *Conférence Internationale pour l'Etude des Epizooties 1921*, Annexe, Statuts, 3 (Art. 4).

97 République Française, Ministère de l'Agriculture, *Conférence Internationale pour l'Etude des Epizooties 1921*, Annexe, Statuts, 3–4 (Art. 6).

98 République Française, Ministère de l'Agriculture, *Conférence Internationale pour l'Etude des Epizooties 1921*, Annexe, Statuts, 4 (Art. 10).

99 République Française, Ministère de l'Agriculture, *Conférence Internationale pour l'Etude des Epizooties 1921*, Annexe, Statuts, 5 (Art. 17).

100 Foreign Office (United Kingdom), George Villiers to His Excellency le Comte de Saint-Aulaire (Ambassador of France, London), June 2, 1922. ADLC: Unions, 429QO/332 (Office International des Epizooties, 1920–23).

101 [Ministère des Affaires Etrangères (France)], Le Président du Conseil, Ministre des Affaires Etrangères à Monsieur l'Ambassadeur de France à Londres, July 19, 1922. ADLC: Unions, 429QO/332 (Office International des Epizooties, 1920–23).

102 [Ministère des Affaires Etrangères (France)], Le Président du Conseil, Ministre des Affaires Etrangères à Monsieur le Ministre de l'Agriculture, November 13, 1922, 3. ADLC: Unions, 429QO/332 (Office International des Epizooties, 1920–23).

103 Ministère des Affaires Etrangères (France), Le Président du Conseil, Ministre des Affaires Etrangères, Letter drafts to embassies (signed Daeschner), June 8, 1923. ADLC: Unions, 429QO/332 (Office International des Epizooties, 1920–23).

104 République Française, Ministère de l'Agriculture, *Conférence Internationale pour l'Etude des Epizooties 1921*, Annexe, Statuts, 4–5 (Art. 11).

105 Légation de la République Française en Autriche, M. Lefevre Pontalis, Ministre de la République Française en Autriche, à Son Excellence M. Poincaré, Président du Conseil, Ministre des Affaires Etrangères (France), July 31, 1922, attached: Note verbale, July 27, 1922. ADLC: Unions, 429QO/332 (Office International des Epizooties, 1920–23).

106 Légation de France en Haiti, M. L. Agel, Ministre de France en Haiti, à Son Excellence Monsieur Poincaré, Président du Conseil, Ministre des Affaires Etrangères (France), July 20, 1922. ADLC: Unions, 429QO/332 (Office International des Epizooties, 1920–23).

107 Légation de la République Française en Equateur, Le Ministre Résident de France en Equateur à Son Excellence le Président du Conseil, Ministre des Affaires Etrangères, January 20, 1923. ADLC: Unions, 429O/332 (Office International des Epizooties, 1920–23).

108 For the example of Syria and Libanon: Haut Commissariat de la République Française au Syrie et au Liban, Le Général Sarrail, Haut Commissaire, à Son Excellence Monsieur le Ministre des Affaires Etrangères, October 3, 1925. ADLC: Unions, 429QO/333 (Office International des Epizooties, 1924–25).

109 [Ministère des Affaires Etrangères (France)], Le Président du Conseil, Ministre des Affaires Etrangères, à Monsieur le Ministre de l'Agriculture. July 4, 1922. ADLC: Unions, 429QO/332 (Office International des Epizooties, 1920–23).

110 Ambassade de l'Union des Républiques Soviétiques Socialistes à Paris, Note verbale, December 30, 1924. ADLC: Unions, 429QO/333 (Office International des Epizooties, 1924–25); [Ministère des Affaires Etrangères (France)], Le Président du Conseil, Ministre des Affaires Etrangères, à Monsieur le Ministre de l'Agriculture, January 15, 1925. ADLC: Unions, 429QO/333 (Office International des Epizooties, 1924–25).

111 World Organisation for Animal Health (OIE), 'International Agreement for the Creation of an Office International des Epizooties in Paris, done in Paris, January 25, 1924', accessed May 15, 2021, https://www.oie.int/about-us/key-texts/basic-texts/international-agreement-for-the-creation-of-an-office-international-des-epizooties/. See also: McVety, *The Rinderpest Campaigns*, 33–4.

112 See also Chapter 6.

113 [Ministère des Affaires Etrangères (France)] to the Ambassade de Japon, France, March 14, 1929. ADLC: Unions, 429QO/334 (Office International des Epizooties, 1929).

5 The League of Nations and international projects addressing animal diseases

In the 1920s and 1930s, several international organisations began to integrate projects of animal disease control into their agenda. This situation offered new opportunities for activities of international cooperation and the networking of veterinary experts. Especially the creation of the OIE was an important step to substantiate international efforts and networks in the thematic area of animal diseases and epizootics control on a permanent basis. Furthermore, the foundation of the League of Nations and the International Labour Organization (ILO) in Geneva in 1919 opened up new possibilities of international activities also in the field of animal disease control. Both institutions structurally integrated experts in their work, provided new platforms of international networking, and actively supported transnational scientific exchange. Geneva became a magnet for the activities of professional, semi- and non-official international associations and movements.[1] In the 1920s, the ILO and the League of Nations Health Organisation (LNHO) initiated a joint project on anthrax in the context of activities for the protection of workers from occupational diseases. A second project of the LNHO dealt with questions of rabies and secure rabies vaccination. Furthermore, in 1927, the same year the OIE took up its work, the League began to organise expert discussions on an international agreement addressing veterinary issues related to the trade and transit of animals and animal products. The League continued its activities in the thematic area of animal health together with its networks of experts until the outbreak of the war in Europe in 1939.

The League of Nations, the ILO and co-operating organisations considerably expanded approaches of veterinary internationalism especially by focusing on several specific projects on zoonotic diseases and veterinary public health. With their approaches, they further developed and expanded the pre-war strategies by promoting a new awareness of the complexity of animal diseases, their interconnections with other economic, political, scientific and social problem areas, and the blurring boundaries between animal and human health problems. The actors involved used the repertoire of veterinary internationalism and organised international conferences, initiated

DOI: 10.4324/9781003033998-5

projects of international conventions, and worked to improve the transfer of information on specific animal disease topics. As already the experiences of the international veterinary congresses of the 19th century had shown, it proved to be a difficult task to negotiate, establish and put into practice official agreements on the international level that addressed transboundary animal disease problems. The new technical committees working in the international organisations in the interwar period therefore also developed scientific-technical recommendations and guidelines intended as non-binding sets of rules, with the aim to communicate these efforts as widely and globally as possible. These forms of international principles opened up opportunities to maintain discussion networks and to take into consideration the global dimension of the problem. While these projects provided important inputs, the non-binding nature of such international rules and guidelines was, at the same time, a main hindrance for a more concerted international approach towards animal disease problems.

Addressing anthrax through international cooperation

The efforts of the ILO and the League of Nations on issues of animal disease control began with a project dealing with the prevention of anthrax infections in the context of industrial work. The issue of anthrax as an occupational disease was a complex and multi-faceted problem area that manifested on several globally intertwined levels. As an animal disease, anthrax occurred in both wild and domestic animals. The disease was transmitted among animals and to humans, often in context of livestock farming, with one of the main modes of transmission being exposure to the bacterial spores. The spores were very resistant to abatement efforts and could survive in a virulent form within the soil, the dust, the water, the garbage, or in animal products. The trade with animal products spread the disease further around the world. Increasingly, therefore, anthrax also became a problem in context of industrial production and global transportation systems, especially in ports, tanneries and the wool industry, since the spores were found in wool, leather, hides or other animal products, resulting in the name "woolsorters' disease".[2] Researchers tried to develop vaccines and experimented with prevention and treatment. Anthrax vaccines for livestock were in use since the late 19th century. Pasteur and his team were publicly celebrated for the famous successful vaccination of sheep in 1881 in the Pouilly-le-Fort experiment with an attenuated vaccine.[3] With the rising transcontinental and transoceanic trade in wool, leather, skins and hides for various industrial purposes, the problem of anthrax gained the attention of governments as a disease that could impair the health of workers in several branches. Industrial and occupational hygiene was gaining in significance in the late 19th century and the first decades of the 20th century as a specialised field that operated across different scientific and medical disciplines. The problem of anthrax infections at workplaces became part of these activities aiming to improve the safety of workers.[4] Anthrax was a specific case in the discussions on workplace safety since it appeared not only in context of industrial and rural labour but also in nature. Before the First World War, the question of human anthrax infections in occupational contexts had already been debated on several international platforms. The series of three International Congresses on Occupational Diseases, of which the first one was held in 1906 in context of the Milan World Exposition, integrated the issue of anthrax as an occupational disease in its discussions.[5] At the congress of 1906, the Permanent International Commission for the Study of Occupational Diseases (Commission Internationale Permanente pour l'Etude des Maladies du Travail) was founded with its seat in Milan. It cooperated with the specialised Milan Labour Clinic (Clinica del Lavoro), inaugurated in 1910, which became a pioneer in the treatment, study and prevention of occupational diseases.[6] In the case of anthrax, their work concentrated on organising educational campaigns about sanitary measures to prevent anthrax infections in factories.[7] In 1912, the International Association for Labour Legislation, a professional organisation with several national branches, created a subcommittee to investigate the problem of anthrax in industrial contexts. It elaborated a detailed list of recommendations for preventive on-site measures such as improving workplace hygiene, disinfecting material of animal origin and workplaces, educating employers and workers, and introducing a regular sanitary inspection of factories by experts. The subcommittee also demanded the introduction of international practices to advance the fight against the disease, such as the elaboration of detailed statistics of the disease for each country, as well as unified sanitary regulations in industrial contexts.[8]

Shortly after the First World War, the International Labour Conference, the plenary assembly of the newly founded ILO, took up the issue of anthrax as an occupational disease in its first meeting in Washington in autumn 1919. The foundation of the ILO resulted from part XIII of the Treaty of Versailles, which mentioned in its preamble the advancement of social justice as a means of securing peace. This included "the protection of the worker against sickness, disease and injury arising out of his employment".[9] At the Washington conference, the question of anthrax emerged in debates on protecting workers, especially women and children, in industrial employment. The conference's Commission of Unhealthy Processes recommended that "international action be taken as the only means of preventing the spread of this disease from animals to man",[10] and suggested an international convention on the matter. The focus should be on infections in the wool industry, as there existed established disinfection methods for this product.[11] The conference adopted a recommendation on the prevention of anthrax transmission through wool as part of a list of six issues for creating safer workplaces. The recommendation advised the members of the conference to establish arrangements for the disinfection of wool either in the exporting country or in the ports.[12] The International Labour Conference of 1919 thus established the problem of anthrax as an occupational disease as a topic for future international debates.

At further sessions of the International Labour Conference, especially the one in 1921 in Geneva, the problem of anthrax infections in occupational contexts was again on the agenda. In preparation for the Geneva meeting, the organisers at the International Labour Office, the secretariat of the ILO, sent a detailed questionnaire about the problem of anthrax in the wool industry to the member governments. The aim was to produce a draft convention.[13] In the accompanying introductory report, the questionnaire interpreted the prevention of anthrax transmission through wool as a model case revealing the need for international cooperation on the issue of workplace safety:

> This problem offers, indeed, a peculiarly suitable field for international co-operation, whether on the question of principle or on the means of applying it. Most civilized countries have done something to prevent anthrax, either in agriculture or in industry, or both; but they have done nothing to control the export and import of dangerous materials.... By international agreement it would be possible to ensure early and efficient disinfection, before export, in the producing countries themselves, since it would be to no one's interest to attempt to evade the regulations, because all fears of the economic inequalities would be removed.... To be effective, this co-operation must be complete and constant. Absolute co-ordination of effort, both as regards disinfection and supervision, is essential. Experience shows that measures adopted by individual states are ineffective.[14]

The problem, as outlined here, was not at all new—the cross-border spread of anthrax spores caused dangers for workers in exporting and transit countries during the processes of production and transportation, and in importing countries on arrival and during industrial processing of commodities. International activities would therefore have to address also the interstitial space between the countries where the pathogens moved around. The idea was again to use the level of 'organising trust'—to work on internationally applicable standards of disinfection and veterinary supervision already in the livestock-producing countries, thereby preventing the problem from acquiring a cross-border dimension. The organisers further justified their discussion on anthrax in industrial contexts by explaining that the disease could cause severe public panics, and emphasised that national measures alone were insufficient.[15] The preparation efforts on the issue of anthrax prevention in the wool industry for the Geneva Labour Conference of 1921 included, next to encouraging activities of national regulation, also the idea of a new international organisation for the prevention of anthrax. This new institution would have, as an international control authority, the task of setting up internationally uniform disinfection procedures and carrying out control measures at ports.[16] Such a suggested international activity for the prevention of anthrax would have exceeded the role of the new OIE which was planned in the same year. The preparations followed the work of the national committee on anthrax of the British government, which had submitted its report in 1918.[17] The British committee of experts, headed by the Liberal parliamentarian Sir William Middlebrook, had conducted an investigation on the disease as an industrial and trading problem, including a special examination on the problem of disinfection of animal products, together with a series of laboratory experiments. They were searching for trustworthy technical methods of disinfection, which would be bacteriologically and chemically safe for the workers and applicable for industrial purposes.[18] The ILO preparatory paper for the 1921 conference considered the results of the British investigations as a basis for suggesting the concept of compulsory disinfection to the governments.[19] The idea of introducing international control measures of compulsory disinfection of wool and animal hairs, however, was not without problems. In particular, there was the question where the procedures of disinfection, which could potentially also have dangerous effects on human health, should be conducted, whether in the exporting or the importing country. One idea was to establish central disinfection stations with internationally standardised procedures in the ports of embarkation, to reduce the risks for those occupied in the transport of animal products.[20]

The third annual session of the International Labour Conference in Geneva took place in November 1921, with the issue of anthrax prevention on the agenda. The conference installed a Commission on Anthrax. It was constituted according to the tripartite organisation of the ILO, assembling in total 30 representatives of governments, employers and workers, among the latter two women.[21] The commission agreed that an international activity to address the problem of anthrax as an occupational disease transmitted through wool was necessary. The discussions highlighted not only the aspect of improving the health of workers, but also the economic arguments to underline the need for an international approach—since national authorities organised the procedures of disinfection (in case they did), a distortion of the wool market would be a possible consequence. With increasing national regulation and control measures, the market in wools would shift; products without disinfection would be traded to countries without the respective import regulations.[22] The commission members concluded that the issue was a delicate one with regard to international economic relations, and that they needed further detailed investigations before any international agreement on anthrax prevention in the case of wool could be envisaged.[23] The conference followed the advice of the anthrax commission and issued a resolution emphasising the aim of a complete eradication of the disease among animals. How that should actually be accomplished required further examination. The conference chose one of the common procedures in these situations of uncertainty: it transferred the problem to a new committee of experts.[24]

As a result, an ILO Advisory Committee on Anthrax convened in December 1922 in London. Most participants were ILO experts or representatives from countries with an important wool production or manufacturing sector: Australia, Belgium, France, Germany, Great Britain, British India, Italy, Japan, Spain, South Africa, and the United States.[25] The head of the committee was Sir William Middlebrook, who had already chaired the British committee on anthrax. The Advisory Committee on Anthrax was charged with debating the infection, transmission and prevention of the disease in different contexts and concerning animal wool and hairs, bones, horns, hoofs, skins and hides, and the prevention of animal infection among flocks.[26] The committee produced a number of proposals, agreeing on the international significance of the problem and advocating in general further steps towards an international agreement.[27] It recommended technical and hygienic procedures and regulations, focusing on the principle of compulsory disinfection of the different animal products. In the case of animal hides and skins, which in most countries played an important role in the transmission of the disease, the committee considered compulsory disinfection as technically difficult to be transacted on a broad scale.[28] Also in case of wool, the procedures of disinfection in the industry were controversially discussed. The British representative of the Indian government opposed a compulsory disinfection of wool as too inflexible, arguing that the numbers of disease cases in many countries were too low to justify such strict and trade-hampering measures.[29] The problem of anthrax thus revealed more and more levels of diversity and complexity. The discussed products of animal origin—wool, hairs, skins, hides, bones, horns, hoofs etc.—reacted differently to procedures of chemical and mechanical disinfection. This created different problems, concerning the safety of workers in the production, processing and transport industries, and also in relation to the economic value of commodities. The Advisory Committee discussed to recommend a differentiation of the question by including other international bodies, especially since the ILO did not have the funds to finance detailed further studies.[30] First, the question of anthrax in rural areas and the protection of agricultural workers was acknowledged as a problem that needed to be tackled, and it was passed on for the attention of the International Institute of Agriculture (IIA) in Rome.[31] Second, the question of anthrax as an occupational disease in industrial contexts, especially with regard to the transmission via hides and skins, should in future be addressed in cooperation with the League of Nations. Luigi Carozzi, the head of the ILO Industrial Hygiene section, argued that the League was the international body with the necessary funds for such detailed and financially precarious inquiries.[32] Indeed, this recommendation was an important step since it reacted to the multifaceted nature of the disease and enlarged the circle of international organisations occupied with the problem area.

The IIA took up the task of investigating anthrax in livestock, even though epizootics and animal diseases had so far not figured among its main lines of activities.[33] During the 1920s, the IIA was an Italian-dominated institution, and agrarian internationalism became intertwined with fascist views on the importance of agriculture and rural society for national development.[34] The transfer of the issue of anthrax in context of agricultural work to the IIA made sense, since it concerned the health of agricultural workers or other persons in contact with animals in rural contexts. In 1924, the General Assembly of the IIA approved the suggestion to conduct an international enquiry on the prevention of anthrax in livestock.[35] The ILO Advisory Committee on Anthrax had already recommended a series of general measures of disease prevention—the mandatory notification of cases of anthrax, isolation of animals suspected of infection, the professional removal of carcasses, and preventive vaccination of livestock.[36] Following from this starting point, the IIA's Bureau of Agricultural Science began working on a new anthrax questionnaire to assess the global disease situation in rural contexts. The questions were to be sent to all official contacts of the IIA as well as to non-member states in order to get a worldwide overview of the disease situation. The IIA aimed to collect the following information: geographical distribution and species of livestock; the state of disease statistics; the activities and results of prophylaxis such as vaccination; the state and results of administrative control; the state of "propaganda, or any other popular measure of a general character"; and the interest of the respective government in an international agreement on the fight against anthrax.[37]

The outcome was a detailed survey, entitled *The Prevention of Anthrax Infection among Live Stock. Results of an International Enquiry*, published in 1925, which covered the worldwide incidences of diagnosed cases of anthrax in context of livestock husbandry and rural work.[38] The global approach of the report turned out to be a chance for the administrations of smaller countries and their veterinary experts to get their voices heard at the international level and state their opinion on an issue of importance for their agricultural economy and public health. The IIA included an evaluation of opinions towards the subject of a possible future international arrangement on anthrax control, which concluded that the majority of national veterinary administrations wished for the establishment of a form of official international cooperation.[39] This catalogue of suggestions from the different national veterinary and sanitary services provided some ideas for new forms of international cooperation on transboundary animal diseases. Most responses focused on the importance of internationally unified legislation of disease control which should prevent the export of diseased livestock. Other suggestions, especially those coming from smaller, less industrialised countries or colonies, included a number of innovative and collaborative ways of approaching the disease. The experts of the French colonial veterinary services of Madagascar, where the disease was a severe problem, stated that an international agreement would not help much to address the anthrax situation on the island. They suggested instead going beyond the usual focus on trade-related issues by conceptualising the containment of anthrax in areas where the disease was enzootic. According to the statement, Madagascar had achieved good results with its programme of livestock vaccination in cooperation with the Pasteur Institute in Antananarivo. They had educated a group of African vaccinators who regularly toured the country. Though there was, as they stated, some resistance from livestock owners, the results of this programme of vaccination were convincing, the report stated.[40] They suggested that a mandatory programme of vaccination, especially where the disease appeared as enzootic, would tackle the problem at its roots on a global scale and consider its dimensions beyond trade-related questions.[41] The experts of the Dominican Republic too suggested a more thorough internationalisation of the fight against anthrax, which should in their opinion start with a uniform and centralised collection of data. An international committee on anthrax, according to this idea, could tour the world, visit all veterinary services and assess the disease situation through internationally uniform statistical methods, and thereby launch worldwide attention to the disease through "anti-anthrax-propaganda".[42] The veterinary services of other countries, such as for instance Uruguay and Czechoslovakia, likewise supported the idea of a more globally oriented campaign against the disease, the experts of the latter country also highlighting the idea of an international inspection service.[43] The report was submitted to the IIA-ILO Advisory Agricultural Committee in November 1925. The IIA announced its decision to leave the future coordination of international control of anthrax to the newly founded OIE, which would take up its work two years later. On this occasion, the IIA expressed its hopes for close cooperation with the new international office.[44]

With regard to the problem of anthrax transmission in the wool industry and the question of disinfection methods of wool and hairs, the International Labour Conference decided in its sixth annual session in May 1924 to not pursue an international activity with regard to the idea of obligatory disinfection.[45] However, to further approach the issue of anthrax transmission via hides and skins on an international scale, the ILO Advisory Committee on Anthrax in its meetings in December 1922 had suggested a joint project of the ILO and the League of Nations.[46] Since many anthrax infections occurred via skins and hides, the question of anthrax transmission through these products was important, especially since no convincing procedure of disinfection was yet established.[47] Furthermore, the anthrax problem in the tannery industry included a dangerous environmental dimension that united industrial and agricultural areas of interest—through the water used in the tanneries, the spores were washed into the fields and could potentially again be distributed by animals, not only by livestock but also by birds and snails.[48] The ILO did not have the technical facilities for further inquiries, therefore it outsourced the investigation. Hermann Leymann, the representative of the German government at the ILO and an expert on occupational safety, compiled a report on anthrax in the tannery industry and on possible methods of disinfection of skins and hides.[49] Leymann's report presented different methods of chemical and thermochemical disinfection. Based on experiences in Germany, he showed that the globalised trade in animal products resulted in a circle of anthrax infections despite national veterinary regulations, which were not sufficient as prevention measures.[50]

In late summer 1923, the ILO's director Albert Thomas officially consulted with the League Secretariat for possible cooperation on the issue of anthrax in hides and skins and the investigation into the means of disinfection as a measure of prevention.[51] Anthrax was one of several cooperative projects.[52] During the early 1920s, the LNHO had already expanded its activities in several areas of international health, with commissions of experts working on an increasing agenda of health concerns; the organisation was an important hub within a network of institutions active in public health.[53] Since 1921, its Epidemiological Intelligence Service collected, centralised and disseminated information on the spread of various diseases with epidemic potential, starting with Eastern Europe and expanding these activities until 1939.[54] In spring 1925, the LNHO founded its Eastern Bureau in Singapore to further enlarge its activities of epidemiological intelligence.[55] The LNHO sought to pursue activities in the field of international health that went far beyond the activities of epidemics control and surveillance. At the same time, the OIHP in Paris was organising the International Sanitary Convention of 1926.[56] The two organisations collaborated in epidemiological intelligence as well as institutionally, with the OIHP Permanent Committee operating as General Advisory Council of the LNHO.[57] In September 1923, the ILO Correspondence Committee on Industrial Hygiene agreed on working out a common research plan together with the League's Health Committee.[58] Ludwik Rajchman, the medical director of the LNHO, received a report by Luigi Carozzi of the ILO's Industrial Hygiene Service and the League Health Section member Hector Rulot.[59] Rajchman, though agreeing on the necessity of investigating anthrax, was sceptical about the ILO's proposal for a new mixed subcommittee and contacted the president of the League's Health Committee, the Danish bacteriologist Thorvald Madsen, director of the Statens Serum Institut in Copenhagen.[60] Rajchman wondered whether it would be better and less costly to appoint only one renowned bacteriologist, perhaps from Germany where numerous scientific research projects on anthrax were conducted.[61] Earlier the same year, Madsen and Rajchman had already acted in a similar manner when entrusting the LNHO project for the international standardisation of non-serological biological products (such as vitamins and hormones) to the British physiologist Henry Hallett Dale.[62] Rulot, who prepared the issue for Rajchman, pointed out the necessity to combine bacteriological research with the question of industrial applicability. A committee of experts, he thought, could establish better contacts with the industry in order to conduct the envisaged work in its scientific-technical and industrial dimensions.[63] Eventually, the Health Committee agreed to a mixed subcommitee of six experts, three from each cooperating organisation.[64] Its task was to study the problem of anthrax in the hides and skins industries, mainly to investigate the research conducted in a number of countries about the suitable and safe methods of disinfection. The mixed subcommitee should collect the results and evaluate the application of the methods in industrial contexts in order to advance the protection of workers.[65] Converting this ambitious plan into an international project of anthrax control with practical consequences turned out to be difficult. The six participants of the mixed subcommitee met in September 1924. They were renowned experts on epidemiology or occupational hygiene: Reynaud (probably Lucien Raynaud, French malariologist and Inspector-General of the Algerian Sanitary Services), Bernhard Nocht (German epidemiologist and specialist in tropical medicine), and Donato Ottolenghi (Italian physician and epidemiologist) on behalf of the League of Nations (all members of the League of Nations Malaria Commission); Désiré Glibert (Belgian chief sanitary inspector), Giovanni Loriga (Italian epidemiologist), and Edgar Leigh Collis (British specialist on industrial hygiene) on behalf of the ILO.[66] Since recent research had produced the approach of using toxic gases as disinfectants against anthrax spores, the subcommitee decided that each member should collect information on new practices with regard to the disinfection of hides and skins in his respective homeland. Ottolenghi, Collis and Nocht should conduct comparative research on chemical methods of disinfection.[67]

In 1925, the ILO took up anthrax as an occupational disease in the Workmen's Compensation (Occupational Diseases) Convention. Together with poisoning by lead and by mercury, anthrax was defined as a disease for which the compensation of workers was included, in accordance with the respective national legislation.[68] The discussion on the convention at the Seventh International Labour Conference in Geneva in May 1925 focused around the question whether occupational diseases such as anthrax should be regarded as equivalent to industrial accidents with regard to compensation.[69] Including occupational diseases in compensation was as such already a difficult issue, since many diseases developed only as a long-term consequence of working conditions.[70] In this context, the case of anthrax, however, was legally a comparatively easy question, since the symptoms of infection usually appeared soon after contact with the pathogen. Even though the disease also occurred in nature, the link to the workplace was in most cases evident, especially in industrial contexts, and the categorisation as an occupational disease acknowledged.[71] The convention included the compensation of anthrax infections in rural-agricultural work, in industrial work, in working with carcasses and other animal residues, and in the transport industry.[72] Through the initiative of the international projects and the ILO list of occupational diseases, there was by the mid-1920s increasing attention towards the problem of anthrax for both industrial and rural workers on the international level, and by mid-1933, 24 countries had ratified the convention.[73]

In the meantime, the initiative of the LNHO and ILO's mixed subcommitee to produce an internationally applicable technical-scientific method of anthrax prevention in the hides and skins industry encountered difficulties. One of the technical problems was how to actually organise large-scale and safe chemical disinfection procedures of animal products which promised effectiveness in industrial practice. Despite scientific research in various countries, the search of the subcommitee members for methods of disinfection of hides and skins did not lead to convincing results that promised to be of practical use under different industrial conditions.[74] By 1929, the subcommitee had assessed a number of research results, but large-scale experiments in cooperation with the tannery industry were not financially and logistically viable; the funding of the League Health Section and the ILO for the project were too limited.[75] The cooperation therefore did, for the moment, not result in a joint LNHO-ILO publication on the issue of anthrax disinfection of hides and skins. The ILO Correspondence Committee on Industrial Hygiene, however, wanted to take further steps and therefore recommended to establish draft regulations.[76] As a consequence, the Industrial Hygiene Service prepared draft regulations for the protection of workers occupied with hides and skins in industrial work. In early 1931, the draft regulations were submitted by the ILO's governing body to the member governments.[77] Since the methods of disinfection of hides and skins were, as the investigations had shown, not satisfactory, the draft regulations concentrated on the recommendation of hygiene measures at workplaces, medical supervision of workers, systematic notification of cases, and several other methods of personal hygiene and disinfection.[78] The regulation text was intended as a non-committal orientation for governments:

> It will thus be seen that, without requiring any profound change in existing regulations, the draft prepared by the Correspondence Committee on Industrial Hygiene could promote a certain unification in the measures already taken and provide useful indication for countries which have not so far adopted any regulation on the matter.[79]

Through these activities, anthrax was classified internationally as an occupational disease and so far listed as the only infectious disease in this context, mainly because of its industrial and trade-related significance. In the ILO and LNHO discussions, the problem of anthrax as an enzootic disease in rural areas and the consequences for workers occupied in the animal husbandry and rural pasture economy was recognised, but it was not at the fore of discussions. Anthrax as a problem of several industrial branches and of transportation systems gained the most international attention for the safety of workplaces. The IIA's inquiry of 1925 showed that already the assessment of information about a complex, multidimensional problem such as anthrax infections in livestock and in rural contexts was difficult on a global scale because of differences in the statistical methods between the contributing national and colonial administrations.[80] The several connected international projects on anthrax containment in the interwar period thus revealed the difficulties in developing internationally compatible and applicable prevention measures for a multidimensional zoonotic disease problem.[81] Suggestions for a more comprehensive international approach, for example through an international vaccination campaign of livestock, a new international convention with binding character, or a new international institution, remained marginal in the discussions. The result was that, in many parts of the world, anthrax continued to appear as a serious occupational hazard: after the Second World War, and despite national programmes of livestock vaccination, anthrax still caused problems in both rural and industrial contexts, as is documented for example in the case of the United States.[82]

Rabies and the search for secure vaccination

In the mid-1920s, another zoonotic disease relevant in global contexts was put on the agenda of the LNHO. Rabies was a disease occurring worldwide among wild and domesticated animals, and it was therefore a particular threat to persons working in rural contexts. Since the transmission of rabies from animals to humans was immensely feared, the presentation of a successful vaccination for animals and for humans by Pasteur and his team in 1885 opened new opportunities of disease treatment and prevention. In the following decades, researchers worked to further develop and improve rabies vaccination, but the different vaccines also included health risks to humans. In 1903, Paul Remlinger and Rifat Bey at the Pasteur Institute in Constantinople experimentally confirmed the viral nature of rabies.[83] The uncertainties concerning rabies and the vaccination methods against the disease were the starting point for a new international project on rabies in the interwar years. Several researchers approached international institutions with the request of efforts against rabies on the international level, above all with view to the topic of safe vaccination. There were mainly two reasons for their aims to initiate an international approach to the disease. First, the general state of information about the distribution of rabies (both animal and human cases) and about the situation of rabies vaccinations on a worldwide scale was deficient. Second, the effectiveness and safety of the different rabies vaccination types for humans was a controversial issue, in particular with regard to the question of post-vaccination paralysis. The researchers suggested that information on the worldwide situation of rabies vaccination should be collected systematically on a large scale, centralised and made available. By reaching out to the League, they hoped to advance an international approach to the problems of rabies in animals and humans on several levels—raising awareness to the problem of rabies and rabies vaccination, benefitting from the League's networks of scientific contacts and opportunities for funding, and using its information and documentation infrastructures for a new global approach to the rabies problem in its several dimensions.

In late 1924, the microbiologist and physician Richard Pfeiffer, professor at Breslau University, directly contacted the League of Nations about the matter of rabies and rabies vaccination.[84] In his letter, he outlined the need to start international cooperation on rabies as an extremely dangerous disease for animals and humans. He urgently called for a new approach of a systematic international collection of information on the methods of vaccination against the disease in human beings and animals, with a special focus on the problem of post-vaccination paralysis of humans. Pfeiffer offered to produce a detailed questionnaire about rabies at his own Institute for Hygiene at Breslau University and to send it to the existing institutes worldwide working on rabies.[85] Pfeiffer's colleague in Breslau, Herbert Lubinski, who was conducting research on cases of paralysis of humans after rabies vaccination, supported the proposal.[86] In the correspondence with the League, it turned out that in 1913 the Pasteur Institute in Paris had already begun with a project to assess information about rabies and rabies treatment on a worldwide scale.[87] The war had interrupted the project and its scientific collaboration network, and in the early 1920s, there was no centrally coordinated, regularly updated information available about rabies in a global context. Many states and colonial administrations had established their own national or regional rabies institutes or departments, some of them well connected with each other, such as the network of Pasteur Institutes. In early 1926, Rudolf Kraus, working at the Serotherapeutisches Institut (Institute for Serotherapy) in Vienna and a specialist on rabies research, sent a request similar to the one of Pfeiffer to the League.[88] Having returned to Vienna after spending the previous decade in Argentina and Chile, Kraus was convinced that initiating international cooperation on the issue of rabies and rabies vaccination was an important issue.[89] Kraus activated two international networks to advance activity on the problem of rabies vaccination. First, he contacted Emile Roux, the director of the Paris Pasteur Institute, in order to inquire on a possible resumption of the Pasteur Institute's pre-war plan to assess global data on rabies. Second, he asked the League of Nations to organise an international conference of rabies experts.[90] Indeed, both Roux and Calmette, the Pasteur Institute deputy director, agreed on the necessity to start international activities on rabies. Calmette suggested to contact the rabies experts Paul Remlinger (of the Tangier Pasteur Institute) and Auguste Marie (affiliated to the Paris Pasteur Institute) to establish a preliminary report on the idea of an international project on rabies vaccination.[91] The project started to take shape. In April 1926, the Health Committee of the LNHO decided on an international rabies conference in cooperation with the Pasteur Institute, intending to assemble the heads of the most important rabies research institutions of the world. In honour of Pasteur, it would take place at his former home in Paris.[92] As Anne Marie Moulin has shown, the Pasteur Institute's international network represented the promise of the universal validity and applicability of biomedical laboratory research produced by a scientific community of experts.[93] With the increasing numbers of institutions and networks across the world, the Pasteur Institutes underlined the tasks of advancing European scientific work also in colonial contexts: Moulin describes the expansion of the institute network as a "clear case of cultural imperialism".[94] Research on rabies and vaccination activities played an important part in their work. For any envisaged international project on rabies with a global scope, the network was necessarily among the most important of partner institutions. The project on rabies vaccination thus showcased the entanglement between imperial research networks with a global outreach and the projects of veterinary and public health internationalism.

The LNHO began preparing the new international approach on the issue of rabies vaccination by collecting information about public and private research institutions that focused on rabies research and treatment. Its "list of the principal anti-rabies institutes"[95] of April 1926 showed how much the question of rabies research, treatment and prevention had acquired worldwide scientific and governmental attention. In Europe, the LNHO listed 50 specialised rabies research institutes (including those in Russia and Turkey). Fifteen institutes were counted in Asia (including Jerusalem), seven in Africa (apart from Cairo all of them in French colonies), eight in the Americas, and one (the Pasteur Institute in Bandung) in "Oceania".[96] Indeed, there was potential for international cooperation if the scientific networks, research collaboration and communication between the institutes were extended. These numbers also revealed that the fight against rabies was a question of national and imperial prestige. With the list at hand, the new project on rabies in its global contexts could be initiated. The LNHO sent an invitation to the national research and sanitary administrations to assign expert delegations to the envisaged First International Conference on Rabies.[97] As Pratik Chakrabarti has shown, the vaccination against the disease, the main subject of the planned international conference, was during the mid-1920s the subject of intensive discourses in the scientific community.[98] There existed by now several vaccine types and methodical procedures, and the issue of their different effectiveness, handling and safety was highly controversial among scientists. The Pasteurian school of researchers attributed the effectiveness of the attenuated Pasteur vaccine, used since the late 1880s, to its 'live' material.[99] Researchers critical of the Pasteur vaccine version argued that the use of the attenuated viral material in the Pasteur vaccine type contributed to its dangerousness. Chakrabarti has described how the vehement scientific debate about 'live' and 'dead' vaccine types created heated controversy, which included the question whether the vaccination with 'live' material could actually trigger so-called laboratory rabies ("rage du laboratoire").[100] To improve the vaccination, in the years before the First World War several institutions specialised on rabies research around the world conducted a series of experiments, especially by treating the viral material with chemicals and with heat. Claudio Fermi in Sassari (Sardinia, Italy) presented a phenolised and heat-treated vaccine type in 1907/08.[101] In 1911, David Semple in Kasauli (India) produced a different phenolised and heat-treated vaccine version based on the infected brain tissue of sheep in which, in contrast to Fermi's version, no 'live' material was traced any more.[102] The Semple vaccine became known as safe and efficient, and it could be better stored and transported, which was important in countries such as India where rabies was a huge problem.[103] In the early and mid-1920s, at the Pasteur Institute in Novi Sad (Kingdom of Serbs, Croates and Slovenes) Adolf Hempt developed a vaccine version based on the treatment of nervous tissue containing viral material with ether. This etherised inactivated vaccine type was applied on humans since 1925. Also the Hempt vaccine was suitable for transportation and could therefore be used for treatments remote from specialised rabies institutes.[104] Further research on the virus and on diverse methods and vaccine types continued in numerous institutions around the world, but many results of experiments included disadvantages in the production, application, storage or transportation of vaccines.

In April 1927, the Conférence Internationale de la Rage (International Rabies Conference) convened in Paris with 59 experts from 26 countries.[105] The conference programme encompassed general questions on rabies and the rabies virus, its emergence and transmission between animals and to humans, as well as on the discussion of rabies prevention and the safety of vaccinations, also with regard to animals.[106] Taking up Pfeiffer's suggestion, the LNHO had distributed in advance a detailed questionnaire in several languages to the anti-rabies institutes worldwide. The responses were intended to give a detailed survey of rabies cases and their treatment with vaccines, the transformation of the virus types through laboratory experimentation, and the experiences with different methods of vaccination.[107] At the conference in Paris, the discussion on the methods of vaccination of humans as conducted at the different institutions was one of the central issues, including the question of serious health effects such as post-vaccination paralysis. The conference participants showed themselves impressed by the research results in British India with regard to the safety of the Semple vaccine.[108] In the resolutions of the conference, the experts confirmed the immunological effectiveness of the Pasteurian method of vaccination as well as of the attenuated and inactivated phenolised or etherised vaccines.[109] With regard to the severe health issues caused in some cases by vaccinations, the conference participants agreed in the resolutions that the etiology of these problems was not entirely clear.[110] To address the disease in a sustainable way in both animals and in humans, the resolutions outlined recommendations for future international anti-rabies work. The participants suggested to organise comparative research in selected research institutes about the effectiveness of inactivated vaccines such as those treated with ether or phenol, as well as research on the different strains of the virus and their immunological reaction.[111] They asked for the commitment of the LNHO on several levels: to explore the possibilities of comparative research on the vaccines and the virus strains, to arrange the distribution of a fixed viral strain suitable for laboratory research, as well as to regularly collect and submit information about the disease to the rabies research institutions worldwide.[112] The resolutions further asked the LNHO to initiate a publication of rabies statistics to ascertain the existence of regularly updated information on the worldwide situation of vaccination, and noted that the Pasteur Institutes needed to provide the statistical research data for this aim.[113] The conference also recommended the regular vaccination of animals, especially dogs, against rabies, and reducing the number of stray dogs to eliminate the virus.[114]

The Health Committee of the LNHO and the Council of the League approved the resolutions of the international rabies conference.[115] The LNHO took up the task of collecting information on rabies vaccination on a global scale and established a standardised questionnaire for assessing information on the situation of rabies and anti-rabies treatment, which was to be distributed to anti-rabies institutes across the world.[116] The aim was to centralise long-term and regular information on the global situation of rabies and of efforts in its treatment and prevention, and to evaluate it in form of statistics. Ideally, such a statistical analysis would not only provide information on the global rabies situation but also lead to conclusions on the safety of the treatment of humans with different vaccine types, as well as information on other factors contributing to rabies in animals and rabies mortality among humans on a worldwide scale. The LNHO thus constituted itself in this project as a hub for the collection of information in a network of anti-rabies institutes. Beginning in 1930, the collection of data on rabies and rabies vaccination resulted in a new publication series, *Analytical Review of Reports from Pasteur Institutes on the Results of Anti-Rabies Treatment*.[117] The next editions of the series were afterwards regularly included in the Bulletin of the LNHO. The British epidemiologist and mathematician Anderson Gray McKendrick, a specialist in medical statistics, became editor of the new series. McKendrick had served from 1914 to 1920 as director of the Pasteur Institute in Kasauli (India) and during the 1920s worked at the Royal College of Physicians in Edinburgh, where he conducted research on statistics of rabies vaccination.[118] The project of collecting and statistically evaluating data on rabies vaccinations during the following years in context of the publication series on rabies offered insights on the state of worldwide rabies research, vaccination practices and public health care. The pattern of analysis was designed to include data on the patients, the context of rabies infection, and the treatment—the vaccine type given after the infection (post-exposure vaccine), the treatment methods, the number of individuals treated, the mortality rate, the transmitting animals, the infection wound, the role of clothing, the location of the wound, the number of days until treatment, and the evidence of post-vaccination paralysis.[119] In the following years, McKendrick received data from a large number of institutions and public health services occupied with rabies vaccination and research. In some cases, the health authorities of states which had become new members of the League contributed data, for example the Soviet Union.[120] Indeed, the project was characterised by a very global outlook because of the many participating research institutions in colonies. Critical questions about the context of the production of the medical-scientific data, especially those contributed by authoritarian states or colonial administrations, were not in the foreground. During the 1930s, the network of contacts of the LNHO with rabies institutes worldwide intensified. When the first edition of the *Analytical Review* was published in 1930, 31 institutes participated in the project, with 31,656 rabies vaccinations of humans.[121] In February 1938, when McKendrick presented his analysis of the data sent to the League until the end of 1935, the data comprised already 758,182 cases of vaccinated patients.[122] Parallel to this considerable enlargement of data, during the mid-1930s the LNHO also discussed a further differentiation of the statistics of rabies vaccination cases in the *Analytical Review*. For the third edition, McKendrick suggested to the LNHO to integrate more information about those who were treated with vaccination and about their side effects. Raymond Gautier of the Health Section of the League Secretariat confirmed that request in February 1933 in a circular letter to the anti-rabies institutions sending their vaccination results:

> Colonel McKendrick, in the Third Analytical Review, expressed the wish that the directors of the institutes carrying out the treatments of subjects belonging to the two ethnic groups (white race and other races) should establish separate forms for Europeans and non-Europeans. I would like to point out the advantages that this approach would bring, even if it might result in a sum of additional work.[123]

In response to this wish, many of the statistics compiled by anti-rabies institutes differentiated from then on between vaccination cases of 'Europeans' and 'non-Europeans'. McKendrick kept the pattern throughout the series. He pointed out methodological problems in this context because in a number of cases, a clear attribution of persons to the respective category could not be made; the case of the inhabitants of Jerusalem, for instance, appeared difficult to him.[124] The result of this differentiation was the general observation of a higher mortality rate among the included 'non-Europeans', even though, as McKendrick outlined in the *Analytical Review* of 1938, the reasons remained statistically unclear.[125]

In May 1934, the Health Section of the LNHO prepared an internal paper evaluating the interim results of its rabies project.[126] The resolutions of the international rabies conference of 1927, the paper summarised, had included several appeals for a new international project on rabies research and treatment. The conference had asked the LNHO to initiate large-scale comparative research concerning the phenolised and etherised vaccine types, as well as on the different virus strains, and it had recommended the centrally organised distribution of a standardised viral type to research institutions for the respective vaccination studies.[127] So far, the paper stated, only the statistical activity, which had been intended as a pilot project, had showed concrete results of the collaboration. Thus, the main success was the publication of a regularly updated overview of data on the different results of anti-rabies vaccination treatments on a worldwide scale.[128] Further, the League had established a new network of communication and information transfer about the disease by being in contact with worldwide anti-rabies institutes. The internal League paper critically pointed out that so far nothing of the original plan, namely to organise concrete large-scale comparative research on rabies with a network of partner research institutions, had been accomplished.[129] The paper suggested to include a Romanian research institute for the planned research on vaccine types and viral strains, since they treated a large number of patients. As a standard virus, the paper recommended the Pasteur virus type of Paris which was already used by the majority of institutions.[130] However, throughout the rest of the decade, there were not many efforts to realise these inputs on the international level. In 1934 and 1935, Claudio Fermi repeatedly urged the LNHO to work on the recommendations of the 1927 international conference: he requested in particular to investigate on the various methods of rabies vaccinations in a comparative study, and submitted a plan of large-scale experimental research which he wished to be conducted.[131] As a well-known authority on rabies research and developer of one of the early phenol-treated vaccine types which was still much in use, Fermi's requests could hardly be ignored.[132] In October 1935, finally, the Health Committee of the LNHO put the question of rabies vaccination back on its agenda. Most of the experts contacted by the LNHO about the vaccine question advised against a new large-scale research project on vaccines because of financial constraints and uncertainties in the scientific outcome.[133] In spring 1937 therefore, the Health Committee rejected Fermi's plans of a new international rabies project for the moment, while tactfully acknowledging the significance of his request.[134] In May 1938, the League sponsored a regional conference on rabies in Central and Eastern European states in Bucharest. The participants stressed the necessity of a second and more extended international meeting to pick up on the issue in the future.[135] But only in 1945 was the project on rabies briefly revived. In May 1945, after the end of the war in Europe, Remlinger, still director of the Pasteur Institute in Tangier, contacted the LNHO to ask about the possibilities of a new international conference on rabies.[136] Gautier, since 1939 the successor of Rajchman as medical director of the LNHO, answered that the League's future was too much at stake to advance any such project at this point of time.[137] In November 1945, however, Gautier initiated the continuation of the worldwide statistical study on rabies by contacting the British medical statistician Major Greenwood, a specialist in the field of epidemiological statistics, asking him to conclude the project of McKendrick, who had died in 1943.[138] Gautier thought it was time to analyse the data material that had been sent to the League during the last years, which had accumulated in total to about 1,270,000 cases of vaccinations.[139] He also intended to bring the statistical rabies project of the LNHO to a scientific conclusion.[140] Greenwood agreed, and in one of its last publications, the LNHO integrated the concluding study on rabies vaccination following McKendrick's statistical patterns.[141] In his analysis, and similar to McKendrick, Greenwood drew the conclusion that the long-term statistical analysis did not show in a scientifically satisfactory way which of the vaccine types were most suitable in rabies treatment; since the differences between the cases were too pronounced, it was not possible to come to a clear scientific result.[142]

The outcome of the long-term study on rabies vaccination, therefore, did not meet the expectations as articulated in the resolutions of the International Rabies Conference of 1927. An international programme of fostering research on and prevention of rabies on a global scale—which would also necessarily need to address the disease in animals, as envisaged at the conference of 1927—proved to be too ambitious and financially costly for the League. The statistical project on rabies research showed the League's role as a hub for networking scientific institutions and researchers with the long-term aim to centralise and evaluate information on a zoonotic disease on a global scale. It also produced a number of insights into the conditions and contingencies of producing scientific research for international projects. The work on rabies on the international level was dependent on the cooperation with national and colonial administrations and research institutions, especially the vaccination programmes set up by European and imperial scientific networks; the cases of rabies treated at the Pasteur Institutes in the colonies produced a large part of the data on the disease and the vaccination. In the long run, however, the small-scale project on rabies, from which contemporary experts had expected far more, was taken up after the Second World War as one of the initiatives of the League and gained a new impetus in the later 20th and the 21st centuries. From 1962 on, the WHO continued the collection of information on rabies in a different format with its *World Surveys of Rabies* (1962–1998) and till today in its data collection on human and animal rabies.[143] Rabies today kills about 60,000 people worldwide per year, which makes the disease a main focus in the cooperative framework on 'one health' of WHO, OIE and FAO.[144] With support of the non-profit organisation Global Alliance on Rabies Control (GARC) and the initiative "United against Rabies", several international activities to contain the disease on a global scale have been started, focusing, among other aspects, on the mass vaccination of dogs. In 2015, the cooperative project "Zero by 30" of the WHO, OIE, FAO and GARC began comprehensive joint activities with the aim to eliminate disease cases in humans completely by 2030.[145]

The League of Nations and projects of international veterinary regulation

From 1927 on, the League of Nations began to address problems of agriculture and animal diseases in context of trade in order to improve international relations by reducing tariffs and other economic barriers. The aim was to avoid conflicts between trading countries regarding animals and animal products, and to balance the idea of unhampered trade with the need to protect livestock herds from diseases.[146] Francis P. Walters, a member of the League Secretariat and after the Second World War the League's chronicler, described in retrospect the League's work in trade issues as significant through "its right of discussion and recommendation":[147] according to Walters, it influenced national economic policies by promoting recommendations through the debates in the Assembly, the projects and the reports of the Economic and Financial Organisation, and the organisation of international conferences on significant economic and trade issues.[148] Patricia Clavin has underlined the significant role of the League in forming an innovative international economic and social agenda in the interwar years, especially during and in the aftermath of the world economic crisis.[149] The League's projects in veterinary-related questions with regard to trade were part of such efforts. In the late 19th and early 20th centuries, the animal disease situation in exporting countries frequently caused import restrictions and conflicts about the justification of sanitary arguments and their possible use for protectionist purposes. At the international veterinary congresses before the First World War, the topic had been regularly discussed: veterinary experts had presented strategies for developing trustful relations between trading partners and more security in the international trade with animals and animal products; governments had approached the issue by concluding bilateral trade agreements.[150]

The World Economic Conference in Geneva in 1927 gave new input for approaching questions of agriculture, food and animal diseases in their global economic dimension.[151] The conference was part of a series of international meetings organised by the League since 1920 to corroborate peaceful international relations through economic stabilisation.[152] In 1925, the Economic Committee started to prepare the conference and included a focus on agriculture in the programme alongside the topic areas of commerce and industry.[153] Delegations from 50 states took part in the conference, including representatives from the United States, the Soviet Union and Turkey, which were not members of the League of Nations at the time.[154] In its non-binding recommendations, the conference members agreed on the necessity to advance international activities against animal and plant diseases to stabilise trade relations. The recommendations mentioned the role of the new OIE for working on the international containment against animal diseases, and underlined that international agreements on sanitary control should especially be careful to avoid forms of "disguised protection".[155] Concrete results in practice in all fields discussed could only be achieved, "if the intentions of the Conference are translated into actions by the Governments", as Hendrikus Colijn, former prime minister of the Netherlands and chairman of the conference's committee on commerce, argued in a review of the conference, adding: "the great need of the present moment is action".[156] In October 1927, the International Conference for the Abolition of Import and Export Prohibition in Geneva also included in its final act a recommendation to the League Council that underlined the urgency to take up common approaches concerning animal and plant diseases through one or several meetings of experts.[157]

One of the results of the conferences of 1927 was that the issue of animal diseases in context of trade relations was from now until the late 1930s on the agenda of the League of Nations.[158] As a result, the League's Economic Committee established until 1939 four subcommitees and committees of experts on several aspects of animal diseases and veterinary public health (a fifth one did not meet anymore), with the aim of reducing trade restrictions and preventing disguised protectionism. In December 1927, the League Council instructed the Economic Committee to set up a meeting of experts to develop a proposal for an international agreement in trade-related veterinary matters. With regard to the area of plant diseases, the League referred to the activities of the IIA and its planned international conference on the matter, which in April 1929 would conclude the International Convention for the Protection of Plants.[159] The first expert group on animal matters, the Sub-Committee of Experts on Veterinary Questions, met between 1928 and 1931 in four sessions.[160] Many of the participating experts were heads of the respective state veterinary services and active on other international platforms, especially as members of the international committee of the new OIE, which was represented by its director Leclainche. Most were Europeans, but gradually the composition became more global (at the first session with experts from Switzerland, Czechoslovakia, Germany, France, Poland, Denmark, Austria; for the second session also from Italy, Great Britain, Yugoslavia, Brazil; for the third session with its focus on meat trade additionally including representatives from New Zealand, Uruguay, Brazil and Argentina).[161] The tasks of the subcommitee included to

> consider what guarantees might be given by countries that export livestock and what facilities might be granted by importing countries on the basis of these guarantees, and to determine in general the best methods for applying veterinary supervision, taking into account the economic interests of the exporting countries and without prejudice to the interests of countries desiring to take precautions against diseases in animals.[162]

The subcommitee referred in its work to the expert discussions at the OIE, which followed the activities by including regular excerpts of its reports and its recommendations in its Bulletin.[163] Cooperating with the LNHO would have been an important input in order to include discussions on veterinary public health in relation to matters of trade. In the first session of the subcommitee in January 1928, Ludwik Rajchman, medical director of the LNHO, visited the meeting. He highlighted the public health dimension of veterinary work and the dangers of diseases transmissible from animals to humans, but at the same time insisted on the differences between public health work and the efforts against epizootics.[164] He pointed out that the essential point was the question of organising trust:

> Trust is at the basis of these efforts; you have to organise it, because there exists a firm principle that also applies in human medicine, namely to use only measures of proven efficacy and to proportion one's rigour towards risks of contagion accordingly.[165]

Rajchman had thereby underlined an essential problem that had occupied veterinary internationalism since the 19th century: how could international approaches to control animal diseases be organised to show effectiveness and be accepted and implemented by national governments—even more so in times of a worldwide economic crisis? And how did the subcommitee interpret the question of what should build up mutual trust in relation to animal disease control, and how trust could be achieved? After receiving the detailed report of the subcommitee to the Economic Committee, completed after the last session of the subcommitee in October 1930, the League Secretariat worked on three draft international conventions.[166] A meeting of governmental representatives in autumn 1934 confirmed the convention texts after revision, and in February 1935 a diplomatic conference met for the signing.[167] The first, the International Convention for the Campaign against Contagious Diseases of Animals, highlighted the necessity of general measures to be undertaken against animal diseases and their spread. As a basis it named the introduction of a state-organised system of professional veterinary control. The convention further included, based on recommendations of the OIE, the principle of regular information concerning the outbreak of certain diseases through an obligatory bimonthly bulletin. Additionally, it outlined the necessity of the immediate telegraphic notification in case of outbreaks of particularly dangerous diseases with an epizootic potential, the possibility of direct contact between state veterinary administrations in case of urgency, as well as the goal of increasing exchange in veterinary research and education.[168] Many of its points thus repeated the demands of veterinary experts outlined and debated at the international congresses of the late 19th century. The second agreement, the International Convention Concerning the Transit of Animals, Meat and Other Products, aimed to facilitate the transit of animals and animal products. It included the conditions concerning the requirements for certificates for animals bound for transit about their origin and health (the text of which was annexed) and the rights of the transit countries concerning border controls, measures of supervision and of stopping the transit. It also included principles on the prevention of diseases in animal transports, especially the requirements for the disinfection of transport wagons.[169] Several points concerning the equipment of the means of transport were included to protect the animals during transit.[170] The third, the International Convention Concerning the Export and Import of Products of Animal Origin (other than Meat), outlined the export and import requirements of specific products of animals (such as hides, skins, bones etc.) under certain conditions. One main condition specified was again the requirements for the certificates of animal origin and health, and the rights of the importing countries in case these health attestations were not satisfactory.[171] From 1936 to 1941, the Soviet Union, Belgium, Latvia, Bulgaria, Turkey and Romania signed and ratified the conventions, which could enter into effect; Iraq and Poland signed and ratified only the first one. As Ursula Reis has outlined, this was an unsatisfactory result after years of efforts in the project: apart from Poland, the governments represented by experts in the League subcommitee did not agree to the conventions.[172] Most of the governments ratifying the conventions were not major powers, or they were newcomers on the international stage. Their participation in the specialised technical international conventions can be interpreted as a governmental strategy of states to be involved more actively in the international system and in international cooperation in general. More important, with regard to the long-term European-dominated discourse on anti-disease measures, signing the conventions also demonstrated an adjustment to European standards of 'civilised' veterinary systems and was thus an option for governments to improve the acceptance of their veterinary systems and enhance their trading opportunities. Today, the three conventions are still part of the UN treaties.[173] After the Second World War, the General Agreement on Tariffs and Trade (GATT) of 1947 took up the topic again and admitted trade barriers if they served to protect animal health, but forms of covert protectionism were not accepted.[174]

The work of the League's subcommitee on veterinary questions had another consequence: topics of animal disease control, veterinary public health and veterinary cooperation in their relation to trade from then on stayed on the agenda of the League throughout the late 1930s. This certainly contributed to the fact that the subject area later also remained in the focus of the United Nations and other international organisations. In summer 1933, the resolutions of the Monetary and Economic Conference in London included a paragraph requesting the Economic Committee to continue its activities of expert discussions on projects of trade-related regulations with regard to the meat trade.[175] The paragraph referred to the International Congress of Refrigeration in Rome in 1928, which had, in cooperation with the IIA, emphasised the necessity to find an internationally unified method of inspection of refrigerated meat.[176] Refrigeration technologies had improved continuously in the first decades of the 20th century, which made frozen meat an increasingly important product on the meat market.[177] In view of the global economic situation, the Economic Committee decided to await a report prepared by the International Institute of Refrigeration (IIR) on the problems of the international meat trade.[178] The IIR, an intergovernmental institute, was established in Paris in 1920 as the successor of a former expert association, the International Association of Refrigeration. Its focus was on the improvement of refrigeration techniques in a global perspective to support commerce. This included the study of international trade of perishable products and finding solutions for trade problems.[179] The investigations of the IIR, of a "purely scientific and technical nature",[180] would, as the Economic Committee hoped, provide a basis for future discussions within the League on the specific topic of the international meat trade. The IIR assembled a committee that included several of the veterinarian experts active on other international platforms; H. C. L. E. Berger, the head of the Dutch veterinary services and at this time also president of the OIE international committee, served as its chair.[181] The composition of the IIR committee also reflected the global and especially transatlantic scope of the international meat trade: three of the eight members came from meat-exporting countries of South America. In July 1934, the IIR committee established a draft of international principles for the inspection of refrigerated meat and submitted the text to the League Economic Committee.[182] Following these preparations, in it session in April/May 1935, at the time the three international veterinary conventions were signed, the Economic Committee thought the circumstances favourable to establish a new subcommitee of experts on the commerce in meat and meat preparations. The working programme centred on the question of how to facilitate trade relations with a view to constructing a possible new international convention.[183] The focus was now on the transboundary, and especially transoceanic, trade and transport of meat: how to agree on and guarantee a set of international regulations concerning the hygiene and slaughter procedures, adequate storage, chemical and technical preparation, packing, labeling and transport of the meat products, as well as transparent documentation and certification of origin and inspection of all meat intended for export.[184] The Economic Committee envisaged a small subcommitee that should represent exporting and importing countries and was globally oriented; suggestions included the United Kingdom, France, Italy, the Netherlands, Switzerland, the Soviet Union, Czechoslovakia, Argentina, the United States, Australia and Uruguay. Two experts would represent the IIR and the OIE.[185] The plan of the Economic Committee and the Secretariat worked well in the beginning. The new subcommitee of experts on export and import of meat and meat preparations met for two sessions, in May 1935 and, for the revision of a draft convention, in April 1938.[186] The worldwide dimension of the meat trade and its interconnections was a central issue in the debates, even though the expected participants from South American meat-producing countries eventually did not attend. But the experts from Canada, Australia and the United States provided a shift toward the commercial interests of non-European countries.[187] In their report to the Economic Committee after the first session, the attending experts cautiously underlined again the technical, non-political and, even more so, the non-economic character of their work:

> The experts, while endeavouring in the light of their personal experience to seek for practical rules calculated to facilitate international trade in the products under consideration, deliberately abstained from examining all questions of an economic character, confining themselves to questions which were exclusively technical.[188]

At the end of the first session, the subcommitee already produced a proposal entitled "Draft Provisions for an International Sanitary Convention on the International Commerce in Meat and Meat Preparations".[189] The participants agreed on which animals were the subject of the convention (cattle, sheep, and hogs), from which parts of the animals (no intestines, brains or other organs), and in what sort of preparation.[190] However, further problems came up: as already in the late 19th century, the differences in the systems of meat inspection—and therefore in the organisation of state veterinary services—were heavily disputed. For the draft convention, the group of experts agreed on establishing a list of basic principles for meat inspection procedures—mandatory ante-mortem and post-mortem inspection of animals, principles for the equipment and hygiene of slaughterhouses and meat-packing buildings, technical details of the cooling instruments and packages in the traffic of meat and specific preservative and colouring additions, as well as the rules to regulate the rejection of meat importations.[191]

When the group of experts met again in April 1938 for the revision of the draft convention text on the issue of meat trade, the political situation and also the composition of the subcommitee had changed. The Italian member Bisanti was no longer taking part, since Italy had left the League in 1937. The British and the U.S. members did not take part either, but this time experts from Argentina, Uruguay and New Zealand joined.[192] This gave the export countries a stronger position, since the most important meat importing countries were absent—Great Britain, the United States, Italy and Germany, which together, as the Canadian expert Young explained, imported about 98 percent of the world's meat.[193] These absences were a major problem for the project. Given previous experiences with the three conventions on veterinary matters, it seemed more and more doubtful that the project would succeed. The report of the subcommitee of experts to the Economic Committee in April 1938 nevertheless defended their work as a contribution to stabilising commerce and therefore international exchange in general:

> They [the participating experts] are aware that for various reasons it [the draft convention] cannot be applied universally, at any rate for the time being. Nevertheless, they feel that the entry into force of the proposed Convention, even between a relatively small number of countries, would be of the greatest value. They feel that it would show the possibility of introducing into and maintenance in the relations between countries importing and exporting meat principles of equity designed to harmonise their relations by reconciling interests which frequently conflict with one another, while giving satisfaction to the legitimate aspirations arising therefrom. It would constitute a kind of international codification of relations in respect of veterinary nutritional hygiene, and the different countries might in any event make use of its provisions when concluding bilateral agreements.[194]

This résumé by the expert group—10 years after the League had begun its projects in veterinary issues to facilitate commercial relations—indicated a change in international strategies in trade-related veterinary issues by highlighting the idea of developing a form of codification for international usage. The recommendation of non-binding measures produced by experts had already been practiced at the international level for decades, such as through the previous resolutions of international veterinary congresses. However, the declared goal had in most cases been an official international agreement on veterinary control measures. International talks between governmental representatives had started in 1872 with the conference on rinderpest in Vienna, where no binding agreement had been passed either. Now, almost seven decades later, and after the League had issued three international conventions, albeit with rather low participation of its member governments, the participating experts emphasised the value of the process of international discussion itself in its informative and documentary power, as well as for orientation in case of bilateral negotiations. In a summary on the process of its work in the diverse fields of veterinary regulation in May 1939, the secretariat of the Economic Committee outlined that it was, for the moment, following the pathway of drafting internationally applicable 'rules' and not pursuing the project of legally binding international conventions:

> ...taking the view that convention has not always proved a successful method of achieving the desired aim, it instructed the experts to draw up 'rules' for the inspection of meat and other measures designed to ensure its wholesomeness. In its report to the Council, the Economic Committee added that Governments might take these rules as a basis when drawing up bilateral commercial agreements, and that, should the idea of multilateral treaties be reverted to later, these rules might perhaps be embodied in them. [195]

The strategy to set aside, if necessary, the idea of negotiating international conventions and instead advocate more flexible regulations, codifications and 'rules' as the basis of veterinary work on a global scale was a pragmatic approach.[196] The Economic Committee thus aimed to support efforts to improve veterinary sanitary regulation and practices in as many countries as possible. Regular exchange between veterinary staff supported, in this thinking, to encourage a better comprehension of the respective scientific and sanitary systems; this in turn should help to decrease misunderstandings in practice, especially between exporting and importing countries.[197] The Economic Committee hoped that establishing non-binding rules for exerting indirect influence on sanitary practices would be more effective on a global level and in times of political and economic crisis, beyond political and ideological boundaries, than pushing forward with the idea of international conventions:

> The adoption of these rules would establish, as between exporting and importing countries, a unity of doctrine and a uniformity of practice which, whilst making due allowance for the need of public health, would have a favourable effect upon international trade.[198]

The experts of the subcommitee on the commerce of meat had recommended a number of additional studies to improve conditions within the international meat trade. The Economic Committee followed this advice. The years 1938 and 1939 saw therefore a further increase of the League's work to facilitate the international trade in animals and animal products. Three more committees of experts were established on internationally controversial aspects: first, a committee for a more detailed enquiry on meat inspection; second, a committee on the trade with living animals; and third, an investigation into problems of meat preservation in international trade. The choice of these thematic areas also reflected a significant trend in international commerce, as preserved products became increasingly important in trade relations. In November 1938, the first one convened—the committee of experts on meat inspection.[199] The requirements compiled by the committee for creating rules of meat inspection read like general instructions on how to create technical information for international usage:

> The experts have endeavoured to give to the rules which they have drawn up a form as clear and precise as possible, in the hope that, whatever language they are translated into, they will retain their clarity and precision. They have added explanatory notes intended to prevent any erroneous interpretations; in some special cases these notes also explain divergent views brought out by the discussion. They have further avoided laying down any requirements that might be regarded as questionable from the scientific standpoint, and they have not dealt with any unnecessary details. It was their desire in this way to facilitate the international trade in meat to the greatest possible extent. Finally, the experts have endeavoured to make the rules as practical as they could, so as to make the greatest possible contribution towards uniformity in national regulations.[200]

A major requirement for the practice of creating non-binding rules, which should be acceptable for both importing and exporting countries, was therefore, according to the report of the committee of experts submitted to the League Economic Committee, that they should be practicable in use and easily translatable into other languages. The report added the "International Rules Relating to the Methods for the Inspection of Meat Intended for the International Trade".[201] The text outlined two pivotal principles of meat inspection that had also been demanded for decades on international platforms: the responsibility of the state and its sanitary authorities for the implementation of rules for meat inspection, and the adequate qualification and education for meat-inspecting staff. Further, it included instructions about the scientific methods of meat inspection as well as professional supervision of processes of animal slaughter and transport.[202]

Regarding the facilitation of exporting and importing living animals, a further new committee of experts produced in only one session in February 1939 an internationally applicable guideline, the "International Rules Concerning the Export and Import of Live Animals".[203] This was remarkable since the subject was always one of the controversial issues on international platforms; living animals were among the most frequent carriers of diseases. But again, the result of the session was the establishment of rules, intended also as an orientation especially for the prospective bilateral or maybe multilateral agreements. The report of the committee of experts fixed the lists of animal species to be of relevance, including even the specialised issues of bee hives and circus animals.[204] It concentrated on details of the conditions of import and the justifications to prohibit importations. As an important requirement of import, it named again the appropriate documentation of the traded animals through certificates of origin and health, referring to the model certificate included in the transit convention of 1935.[205] Further points elaborated on regulating quarantine rules, transport conditions and hygiene, methods of disinfection, and instruments of diagnostics and their appropriate usage in trade and transport.[206]

A further committee of experts was planned on the topic of the preservation of animal products.[207] This was an issue of growing international significance. The topic showed how the fields for international veterinary-related discussions expanded, since an investigation into rules for the preservation of food needed expertise from several scientific fields, especially concerning refrigeration techniques, food hygiene and food chemistry. The trade in preserved meat products, such as tinned meat, was increasing, and there were numerous problems caused by the complexity of preservation processes and by the different systems of national regulation.[208] In May 1939, the Economic Section of the Secretariat began with preparations to define the problem and the scope of work. The expert consulted to advise the Secretariat on the matter was Camille Boussard, head of the veterinary services at the French Ministry of Agriculture and the chairperson of the previous committee on meat inspection. Together with Emile Nicolas, the inspector general of the French veterinary schools, Boussard submitted a scheme of study with general as well as specialised rules of meat preservation that outlined the significance of finding some form of international handling of the issue. Tin as a packaging material, chemical preservation and agglutinating substances, the processes of drying, salting, heating, refrigerating and smoking food products: all these could possibly constitute, or turn into, considerable health hazards for human beings.[209] The discussions of this new committee of experts on meat preservation—enlarged to encompass also fish and seafood—would have focused on a multitude of problems of worldwide trade in preserved food products. The start of the Second World War in Europe, however, stopped the preparations and further meetings of the committee of experts, as well as in general the League's projects in the field of international veterinary regulation and the safety of products of animal origin.

The work of the expert committees in the late 1930s took as a starting point the question of animal disease problems in trade relations, but went beyond this topic in thematic content and results. After the experiences with the difficulties of establishing international conventions on trade-related veterinary regulation, the work of the League of Nations on international veterinary cooperation continued and expanded in the late 1930s.

The issues debated in the field of veterinary public health and hygiene were of an increasingly complex and interdisciplinary character. The planning of the last committee of experts on the topic of meat preservation especially showed an important thematic shift, not only to facilitate the trade with meat products, but also to include questions of consumer protection in their transboundary dimension. The intention to continue the work on the subject area since the mid-1930s, despite the difficulties associated with establishing international veterinary conventions, can certainly be interpreted as part of the League's increasing efforts to foster broad international cooperation on technical issues. The idea of regulating international work through conventions shifted in favour of strategies that seemed to be easier and more pragmatic ways of improving the prevention and spread of animal-related diseases on a more global scale. Propagating non-binding guidelines or rules as an orientation for national veterinary services worldwide called for functioning networks of the scientific community, as well as for close collaboration with other international organisations such as the OIE or the IIR, to further distribute, discuss and update the guidelines. This direction of projects pointed to the overall aim to gradually improve the worldwide level of animal health, food safety and public health practices through a multi-levelled approach of many small steps. The veterinary projects of the League of Nations ended abruptly in 1939, but despite the difficulties, political tensions and non-realised projects, they certainly had an impact. The projects resulted in a more diverse output of pathways and practices of the idea of 'organising trust': the guidelines produced through international endeavours intended, as non-binding instruments, a more long-term—albeit uncertain—impact on improving national veterinary and sanitary systems. Further, the discussions showed that the topic areas of veterinary internationalism gradually started to become thematically more comprehensive than in the decades before. They systematically included topics of veterinary public health and food safety, and focused on the intersections of these areas of growing importance with regard to international trade relations.

Notes

1 Glenda Sluga, *Internationalism in the Age of Nationalism* (Philadelphia: University of Pennsylvania Press, 2013), 45–78; Madeleine Herren, *Internationale Organisationen seit 1865. Eine Globalgeschichte der internationalen Ordnung* (Darmstadt: Wissenschaftliche Buchgesellschaft, 2009), 54–8.

2 Susan D. Jones, *Death in a Small Package. A Short History of Anthrax* (Baltimore: Johns Hopkins University Press, 2010), 82–3; Rosemary Wall, *Bacteria in Britain, 1880–1939* (Abingdon, New York: Routledge, 2016), 65–6.

3 M. Lombard, P. P. Pastoret, and A. M. Moulin, 'A Brief History of Vaccines and Vaccination', *Revue Scientifique et Technique (OIE)* 26, no. 1 (2007): 34. For a summary on historical research on the anthrax bacillus: Peter C. B. Turnbull and Sean Shadony, 'Anthrax from 5000 BC to AD 2010', in *Bacillus Anthracis and Anthrax*, ed. Nicholas H. Bergman (Hoboken, NJ: Wiley-Blackwell, 2011), 1–15.

4 About the legal situation on anthrax in occupational contexts in Europe and the United States in the early 20th century see a study commissioned by the U.S. Bureau of Labor Statistics: John B. Andrews, *Anthrax as an Occupational Disease*, Bulletin of the United States Bureau of Labor Statistics 205 (Washington: Government Printing Office, 1917).

5 Andrews, *Anthrax as an Occupational Disease*, 98.

6 For the Milan Labour Clinic: Luigi Devoto, 'The Milan Clinic for Occupational Diseases. Its Origin, Purposes and Activities', in *Diseases of Occupation and Vocational Hygiene*, ed. George Martin Kober and William C. Hanson (Philadelphia: P. Blakiston, 1916), 765–76; Andrews, *Anthrax as an Occupational Disease*, 97–8.

7 Andrews, *Anthrax as an Occupational Disease*, 97–8.

8 Andrews, *Anthrax as an Occupational Disease*, 98, 121–23. About the work of the International Association for Labour Legislation on anthrax see also: H. J. W. Hetherington, *International Labour Legislation* (London: Methuen, 1920), 8–10; Wall, *Bacteria in Britain*, 112–13.

9 'The Versailles Treaty: Part XIII, Labour', The Avalon Project. Documents in Law, History and Diplomacy, accessed March 19, 2021, http://avalon.law.yale.edu/imt/partxiii.asp (Section I: Organisation of Labour, Preamble). About the ILO's work on occupational health: Thomas Cayet, Paul-André Rosental, and Marie Thébaud-Sorger, 'How International Organisations Compete: Occupational Safety and Health at the ILO, a Diplomacy of Expertise', *Journal of Modern European History* 7, no. 2 (2009): 174–96.

10 League of Nations and International Labour Office, eds., *International Labor Conference. First Annual Meeting, October 29, 1919–November 29, 1919* (Washington: Government Printing Office, 1920), Report of the Commission of Unhealthy Processes, Appendix, 253, also: Anthrax in Britain, Appendix, 253–5.

11 League of Nations and International Labour Office, *International Labor Conference 1919*, Report of the Commission of Unhealthy Processes, Appendix, 253.

12 League of Nations and International Labour Office, *International Labor Conference 1919*, Recommendation Concerning the Prevention of Anthrax, Appendix, 261.

13 International Labour Office, ed., *Disinfection of Wool Infected with Anthrax Spores: Questionnaire* (Geneva, 1920), 12–13.

14 International Labour Office, *Disinfection of Wool*, 27–8.

15 International Labour Office, *Disinfection of Wool*, Introductory Note, 5.

16 International Labour Office, *Disinfection of Wool*, 27–31.

17 International Labour Office, *Disinfection of Wool*, 18–27; Departmental Committee on Anthrax (Great Britain, Home Office), *Report of the Departmental Committee Appointed to Inquire as to Precautions for Preventing Danger of Infection from Anthrax in the Manipulation of Wool, Goat Hair, and Camel Hair* (London: H. M. Stationery Office, 1918).

18 On the work of the British departmental committee of experts and the results see: International Labour Office, *Disinfection of Wool*, 18–19.

19 International Labour Office, *Disinfection of Wool*, 22.

20 International Labour Office, *Disinfection of Wool*, 22–3.

21 Commission du Charbon/Commission on Anthrax, in: League of Nations and International Labour Office, eds., *Conférence Internationale du Travail/International Labour Conference, 3rd Session Geneva 1921* (Geneva: International Labour Office, 1921), LXVIII, 94.

22 Debate about the Report of the Anthrax Commission, 18th Session, November 14, 1921, in: League of Nations and International Labour Office, *International Labour Conference 1921*, 338–40.

23 Report of the Anthrax Commission, in: League of Nations and International Labour Office, *International Labour Conference 1921*, 703–11, esp. 708.
24 Resolution proposed by the Anthrax Commission, in: League of Nations and International Labour Office, *International Labour Conference 1921*, 866–7, 340. See also: International Labour Office, ed., *Advisory Committee on Anthrax, London, December 1922. Report of the Committee* (Geneva: International Labour Office, 1923), 7.
25 International Labour Office, *Report Advisory Committee on Anthrax*, 8.
26 International Labour Office, *Report Advisory Committee on Anthrax*, 7.
27 International Labour Office, *Report Advisory Committee on Anthrax*, 8, 12–13.
28 International Labour Office, *Report Advisory Committee on Anthrax*, 12.
29 Minute of dissent by the representative of the Indian government (Lt. Col. Hutchinson), in: International Labour Office, *Report Advisory Committee on Anthrax*, 84–5.
30 International Labour Office, *Report Advisory Committee on Anthrax*, 56.
31 International Labour Office, *Report Advisory Committee on Anthrax*, 63. See also: International Labour Office, ed., 'Prévention de l'Infection Charbonneuse parmi les Troupeaux', *ILO Official Bulletin* 8 (1923): 104.
32 International Labour Office, *Report Advisory Committee on Anthrax*, 56. About Luigi Carozzi (1875–1963) and his work at ILO: Cayet, Rosental, and Thébaud-Sorger, 'How International Organisations Compete', 175–6.
33 The IIA published an overview of its activities in 1941: International Institute of Agriculture/Institut International d'Agriculture, *Quelques Aspects de l'Activité de l'Institut International d'Agriculture (1905–1940)* (Rome: Institut International d'Agriculture, 1941). For a contemporary analysis of the IIA: Asher Hobson, *The International Institute of Agriculture: An Historical and Critical Analysis of Its Organization, Activities and Policies of Administration* (Berkeley: University of California Press, 1931).
34 Madeleine Herren, 'Fascist Internationalism', in *Internationalisms. A Twentieth-Century History*, ed. Glenda Sluga and Patricia Clavin (Cambridge: Cambridge University Press, 2017), 203–6; Madeleine Herren and Sacha Zala, *Netzwerk Aussenpolitik. Internationale Kongresse und Organisationen als Instrumente der schweizerischen Aussenpolitik 1914–1950* (Zürich: Chronos, 2002), 83–8, 156–65. About the relations between the IIA and the League of Nations: Elisabetta Tollardo, *Fascist Italy and the League of Nations, 1922–1935* (London: Palgrave Macmillan, 2016), 34–7.
35 International Institute of Agriculture/Institut International d'Agriculture, ed., *The Prevention of Anthrax Infection among Live Stock. Results of an International Enquiry* (Rome: Printing Office of the Chamber of Deputies, 1925), 2.
36 International Labour Office, *Report Advisory Committee on Anthrax*, 10.
37 Questionnaire, in: International Institute of Agriculture/Institut International d'Agriculture, *Prevention of Anthrax*, 4.
38 International Institute of Agriculture/Institut International d'Agriculture, *Prevention of Anthrax*, 123.
39 International Institute of Agriculture/Institut International d'Agriculture, *Prevention of Anthrax*, 126.
40 International Institute of Agriculture/Institut International d'Agriculture, *Prevention of Anthrax*, 50–2.
41 International Institute of Agriculture/Institut International d'Agriculture, *Prevention of Anthrax*, 123.
42 International Institute of Agriculture/Institut International d'Agriculture, *Prevention of Anthrax*, 124.
43 International Institute of Agriculture/Institut International d'Agriculture, *Prevention of Anthrax*, 125.

44 International Institute of Agriculture/Institut International d'Agriculture, *Prevention of Anthrax*, 127.
45 League of Nations and International Labour Office, eds., *Conférence Internationale du Travail/International Labour Conference, 6th Session Geneva 1924*, vol. 1 (Geneva: International Labour Office, 1924), 310–39 (vote: 334-337).
46 International Labour Office, *Report Advisory Committee on Anthrax*, 11.
47 International Labour Office, *Report Advisory Committee on Anthrax*, 11; Wall, *Bacteria in Britain*, 113–14.
48 Achille Sclavo, 'Anthrax', in *Occupation and Health. Encyclopaedia of Hygiene, Pathology, and Social Welfare, No. 36*, ed. International Labour Office: ILO Correspondence Committee on Industrial Hygiene, vol. 1 (Geneva: International Labour Office, 1930), 2.
49 Hermann Leymann, *Anthrax in the Tannery Industry*, ed. International Labour Office (Geneva: International Labour Office, 1923).
50 Leymann, *Anthrax in the Tannery Industry*, 9–10.
51 Thomas to the General Secretary of the League of Nations (Drummond), September 7, 1923. LON R902, 12B/30700/30700.
52 About the cooperative projects of ILO and LNHO see: Paul Weindling, 'Social Medicine at the League of Nations Health Organisation and the International Labour Office Compared', in *International Health Organisations and Movements, 1918–1939*, ed. Paul Weindling (Cambridge: Cambridge University Press, 1995), 143–4.
53 For a contemporary summary of the League's health work: Frank G. Boudreau, *Ancient Diseases – Modern Defences. The Work of the Health Organization of the League of Nations* (New York: Columbia University Press, 1939). For a history and synthesis of the work of the LNHO: Iris Borowy, *Coming to Terms with World Health. The League of Nations Health Organisation, 1921–1946* (Frankfurt am Main and New York: Peter Lang, 2009), 18–25.
54 Andrew D. Cliff and Matthew Smallman-Raynor, *Oxford Textbook of Infectious Disease Control. A Geographical Analysis from Medieval Quarantine to Global Eradication* (Oxford: Oxford University Press, 2013), 43.
55 About the LNHO Eastern Bureau: Cliff and Smallman-Raynor, *Infectious Disease Control*, 43–8; Tomoko Akami, 'A Quest to Be Global: The League of Nations Health Organization and Inter-Colonial Regional Governing Agendas of the Far Eastern Association of Tropical Medicine 1910–25', *The International History Review* 38, no. 1 (2016): 1–23.
56 On the International Sanitary Convention of 1926 in a global context: Anne Sealey, 'Globalizing the 1926 International Sanitary Convention', *Journal of Global History* 6, no. 3 (2011): 431–55.
57 For an overview on the relations between the OIHP and the League of Nations Health Organisation: Cliff and Smallman-Raynor, *Infectious Disease Control*, 40–1.
58 International Labour Organization/Organisation Internationale du Travail, 'Première Session de La Commission Consultative d'Hygiène Industrielle', *ILO Bulletin Officiel* 8 (Juillet-Décembre 1923): 134–5. About the ILO's work in industrial hygiene and occupational health see also: Cayet, Rosental, and Thébaud-Sorger, 'How International Organisations Compete'.
59 Thomas to Rajchman, July 31, 1923. LON R902, 12B/30055/9891; Rapport concernant la Collaboration de la Commission Consultative d'Hygiène Industrielle du Bureau International du Travail et le Comité d'Hygiène de la Société des Nations au Sujet de l'Etude du Charbon (Dr. Carozzi et Dr. Rulot), transmis le Novembre 14, 1923. LON R902, 12B/30700/30700.
60 Thorvald Madsen (1870–1957) was director of the Statens Serum Institut in Copenhagen and president of the League Health Committee from 1921 to 1937. About Madsen's activities in international health see: Anne Hardy,

'Actions Not Words. Thorvald Madsen, Denmark, and International Health, 1902–1939', in *Of Medicine and Men. Biographies and Ideas in European Social Medicine between the World Wars*, ed. Iris Borowy and Anne Hardy (Frankfurt am Main: Lang, 2008), 127–43.

61 Rajchman to Madsen, November 19, 1923. LON R902, 12B/32307/30700.

62 Borowy, *Coming to Terms with World Health*, 115. Dale received the Nobel Prize in Physiology or Medicine in 1936 for his studies on neurotransmission.

63 Note à M. le Docteur Rajchman, Directeur Médical, sur le Charbon, par le Dr. H. Rulot, November 26, 1923. LON R902, 12B/30700/30700.

64 LN Health Committee: Report on the Work of the First Session, February 11 to 21, 1924, Anthrax Infection: 6–7. LON R902, 12B/34398/30700.

65 LN Health Committee: Report on the Work of the First Session, February 11 to 21, 1924, Anthrax Infection: 6–7. LON R902, 12B/34398/30700.

66 LNHO, Report of the Temporary Mixed Committee on Anthrax, September 30, 1924, 1. LON R902, 12B/40053/30700.

67 LNHO, Report of the Temporary Mixed Committee on Anthrax, September 30, 1924, 7. LON R902, 12B/40053/30700.

68 Workmen's Compensation (Occupational Diseases) Convention (No. 18), June 1925, Art. 2: List of Diseases and Toxic Substances/List of Corresponding Industries and Processes, in: International Labour Organization, 'C018: ILO Workmen's Compensation (Occupational Diseases) Convention, 1925 (No. 18)', NORMLEX: Information System on International Labour Standards, accessed March 19, 2021, http://www.ilo.org/dyn/normlex/en/f?p=1000:12100:0::NO::P12100_INSTRUMENT_ID:312163.

69 International Labour Office, ed., *International Labour Conference, 7th Session, Geneva 1925: Report on Workmen's Compensation* (Geneva: International Labour Office, 1925), 153–5.

70 International Labour Office, *Report on Workmen's Compensation 1925*, 155.

71 International Labour Office, *Report on Workmen's Compensation 1925*, 157, 183.

72 Workmen's Compensation (Occupational Diseases) Convention (No. 18), June 1925, Art. 2: List of Diseases and Toxic Substances/List of Corresponding Industries and Processes, in: International Labour Organization, 'NORMLEX ILO Workmen's Compensation (Occupational Diseases) Convention 1925'.

73 International Labour Organization, 'NORMLEX ILO Workmen's Compensation (Occupational Diseases) Convention 1925' (ratifications by country).

74 Wall, *Bacteria in Britain*, 115; A. Grut, 'The Work of the International Labour Organization in Occupational Health', *British Journal of Industrial Medicine* 8, no. 4 (1951): 202–3.

75 International Labour Office, Sous-Comité Mixte du Charbon, Réunion des 17 et 18 juillet 1929, June 20, 1929. LON R5880, 8A/12251/3856; Relations with the League of Nations, in: International Labour Organisation, ed., *ILO Yearbook* (Geneva: International Labour Office, 1930), 42. See also: Grut, 'The Work of the International Labour Organization in Occupational Health', 202–3.

76 LN/International Labour Office, Dossier (on draft regulations for protection against anthrax in hides and skins industry), stamped April 23, 1931. ILO Archives: Hy 550/8.

77 International Labour Office: Fifty-First Session of the Governing Body, January 28, 1931, 5th Item on Agenda, plus Draft Regulations for the Protection against Infection by Anthrax in Hides and Skins Industry. ILO Archives: ILO Governing Body sess. 51, D751/100/5. See also: Grut, 'The Work of the International Labour Organization in Occupational Health', 202–3.

78 About the draft regulations especially in context of British measures: Wall, *Bacteria in Britain*, 115.

79 International Labour Office: Fifty-First Session of the Governing Body, Jan. 28, 1931, 5th Item on Agenda, plus Draft Regulations for the Protection against Infection by Anthrax in Hides and Skins Industry, citation p. 6. ILO Archives: ILO Governing Body sess. 51, D751/100/5.

80 On the differences in statistical assessments of animal as well as human anthrax: Sclavo, 'Occupation and Health: Anthrax', 3–5.

81 See also: Grut, 'The Work of the International Labour Organization in Occupational Health', 202.

82 Karl Friedrich Meyer, *The Zoonoses in their Relation to Rural Health* (Berkeley: University of California Press, 1955), 7.

83 B. Melikoglu, R. T. Bagasaç Gül, and T. Özkul, 'Paul Ambroise Remlinger: A Pasteurien in Turkey and His Studies on Rabies', *Revue de Médecine Vétérinaire* 160, no. 7 (2009): 374–7.

84 Pfeiffer to LNHO, December 8, 1924. LON R948,12B/41192/41192. Richard Pfeiffer (1858–1945), since 1909 professor at Breslau University, was renowned for a number of important discoveries in bacteriology and immunology. See: Werner E. Gerabek, 'Richard Pfeiffer', in *Neue Deutsche Biographie*, vol. 20 (Duncker & Humblot: Berlin, 2001), 323.

85 Pfeiffer to LNHO, December 8, 1924. LON R948,12B/41192/41192.

86 Lubinski to LNHO, February 22, 1926. LON R948,12B/41192/41192. Herbert H. Lubinski (1892–?) was head of the Breslau Pasteur Institute until his exclusion by the National Socalist government; for his research on rabies in the mid-1920s: Herbert Lubinski, 'Zur Ätiologie der nach Schutzimpfung gegen Lyssa vorkommenden Lähmungen', *Klinische Wochenschrift* 5, no. 31 (July 1, 1926): 1419–21. For a biographical note: Nathan Koren, *Jewish Physicians: A Biographical Index* (John Wiley and Sons, 1973), 215.

87 Madsen to Pfeiffer, May 27, 1925. LON R948,12B/41192/41192.

88 The Austrian physician and researcher Rudolf Kraus (1868–1932) was since 1923 head of the Vienna Institute for Serotherapy; from 1928 onwards, he again worked in Chile. About Kraus' work in South America: Juliana Manzoni Cavalcanti, 'Rudolf Kraus em busca do "ouro da ciência": a diversidade tropical e a elaboração de novas terapêuticas, 1913–1923', *História, Ciências, Saúde-Manguinhos* 20, no. 1 (March 2013): 221–37. Kraus was co-author of a book on rabies as a zoonotic disease: Rudolf Kraus, Franz Gerlach, and Fritz Schweinburg, *Lyssa bei Mensch und Tier* (Berlin und Wien: Urban & Schwarzenberg, 1926).

89 Kraus to Madsen, February 22, 1926. LON R948, 12B/41192/41192.

90 LNHO, Note on the Convening of an International Conference on Rabies, April 1926 (no day), 1. LON R948, 12B/50886/41192.

91 LNHO, Note on the Convening of an International Conference on Rabies, April 1926 (no day), 2. LON R948, 12B/50886/41192. Paul Remlinger (1871–1964) served from 1911 to 1957 as head of the Institut Pasteur Tangier. For a biographical note: Institut Pasteur, 'Institut Pasteur: Repères Chronologiques. Paul Remlinger (1871–1964)', accessed March 19, 2020, https://web.archive.org/web/20101221105424/http://www.pasteur.fr/infosci/archives/rem0.html. Auguste (Charles) Marie (1864–1935) became in 1933 head of the Institut Pasteur's anti-rabies services. For a biographical note: Institut Pasteur, 'Institut Pasteur: Repères Chronologiques. Auguste-Charles Marie (1864–1935)', accessed April 1, 2021, https://web.archive.org/web/20110823091919/http://www.pasteur.fr/infosci/archives/mca0.html.

92 LNHO, Note on the Convening of an International Conference on Rabies, April 1926 (no day), 2. LON R948, 12B/50886/41192; Madsen to Pfeiffer, May 27, 1926. LON R948, 12B/41192/41192.

93 Anne Marie Moulin, 'The Pasteur Institute's International Network: Scientific Innovation and French Tropisms', in *Transnational Intellectual Networks. Forms of Academic Knowledge and the Search for Cultural Identities*, ed. Christophe Charle, Jürgen Schriewer, and Peter Wagner (Frankfurt am Main and New York: Campus, 2004), 136–7.

94 Moulin, 'The Pasteur Institute's International Network', 136–7, 153, citation: 162.

95 List of the Principal Anti-Rabies Institutes, in: LNHO, Note on the Convening of an International Conference on Rabies, April 1926 (no day). LON R948, 12B/50886/41192.

96 List of the Principal Anti-Rabies Institutes, in: LNHO, Note on the Convening of an International Conference on Rabies, April 1926 (no day). LON R948, 12B/50886/41192.

97 LNHO, Circular Concerning the International Rabies Conference, June 27, 1926. LON R948, 12B/41192/41192.

98 About the scientific discourses on rabies vaccination: Pratik Chakrabarti, '"Living versus Dead": The Pasteurian Paradigm and Imperial Vaccine Research', *Bulletin of the History of Medicine* 84, no. 3 (2010), 387–423.

99 Chakrabarti, 'Living versus Dead', 389–90.

100 Chakrabarti, 'Living versus Dead', 395.

101 R. Barth, H. L. Bock, and G. Korger, 'Geschichte der Tollwutschutzimpfung des Menschen im deutschsprachigen Raum', *Medizinhistorisches Journal* 24, no. 3–4 (1989): 336.

102 Barth, Bock, and Korger, 'Tollwutschutzimpfung', 336–7. In detail about the development of Semple's vaccine version: Chakrabarti, 'Living versus Dead', 392–5.

103 Chakrabarti, 'Living versus Dead', 393–4.

104 Barth, Bock, and Korger, 'Tollwutschutzimpfung', 338.

105 Liste des Participants à la Conférence Internationale de la Rage, in: Auguste C. Marie, Paul Remlinger, and Henri Vallée, eds., 'Conférence Internationale de la Rage', *Annales de l'Institut Pasteur* 42, no. 12, Supplément (Dec. 1928): 166–9. On the conference discussions see: Chakrabarti, 'Living versus Dead', 396–7, 405.

106 Marie, Remlinger, and Vallée, 'Conférence Internationale de la Rage', 5–6.

107 LNHO, Questionnaire Relating to Anti-Rabic Vaccination. LON R948, 12B/52551/41192; LNHO, Réponses aux Questionnaire relatif à la Vaccination Anti-Rabique, February-March 1927. LON 12B/55480/41192.

108 Chakrabarti, 'Living versus Dead', 396–7. See also about the work of the Indian Pasteur Institutes: LNHO, Réponses aux Questionnaire relatif à la Vaccination Anti-Rabique Reçues des Indes Britanniques, Kasauli (Punjab), March 1927. R951, LON 12B/55480/41192.

109 Résolutions, point II/1 and 2, in: Marie, Remlinger, and Vallée, 'Conférence Internationale de la Rage', 7–8.

110 Résolutions, point II/3, in: Marie, Remlinger, and Vallée, 'Conférence Internationale de la Rage', 8.

111 Résolutions, point II/5 and 6, in: Marie, Remlinger, and Vallée, 'Conférence Internationale de la Rage', 8.

112 Résolutions, point II/7 and 8, in: Marie, Remlinger, and Vallée, 'Conférence Internationale de la Rage', 9.

113 Résolutions, point II/10, in: Marie, Remlinger, and Vallée, 'Conférence Internationale de la Rage', 9.

114 Résolutions, point III, in: Marie, Remlinger, and Vallée, 'Conférence Internationale de la Rage', 10.

115 League of Nations Information Section, *Health Organisation* (Geneva: League of Nations, 1931), 24.
116 LNHO, Schedule of Results of Anti-Rabies Treatment, March 28, 1928. LON R5858, 8A/3617/1214.
117 Anderson Gray McKendrick, *Première Revue Analytique des Rapports des Instituts Pasteur sur les Résultats de la Vaccination Antirabique* (Genève: Société des Nations, 1930). Nine further editions were included in the *Bulletin Trimestriel/Bulletin* of the LNHO until 1940.
118 About Anderson Gray McKendrick (1876–1943) and his work in rabies research and statistics: Chakrabarti, 'Living versus Dead', 401.
119 McKendrick, *Première Revue Analytique sur les Résultats de la Vaccination Antirabique.*
120 Traitement Antirabique dans l'U.R.S.S, in: Anderson Gray McKendrick, 'Cinquième Revue Analytique des Rapports des Instituts Pasteur sur les Résultats de la Vaccination Antirabique', *Bulletin Trimestriel de l'Organisation d'Hygiène* 3 (1934): 680–9.
121 LNHO, Note sur la Rage, May 15, 1934, 1. LON R6110, 8A/11384/11384.
122 Anderson Gray McKendrick, 'Huitième Revue Analytique des Rapports des Instituts Pasteur sur les Résultats de la Vaccination Antirabique', *Bulletin de l'Organisation d'Hygiène* 7 (1938): 5.
123 LNHO, Gautier [to the Directors of Anti-Rabies Institutes], February 7, 1933. LON R6066, 8A/1983/1167 (translation: CK). Raymond Gautier (1885–1957) worked since 1924 at the League of Nations Health Section, especially in the area of biological standardisation, and was head of the Eastern Bureau at Singapore from 1926 to 1930. In 1939, he became acting medical director. About Gautier see the obituary in: 'Raymond Gautier, M.D.', *The British Medical Journal* 1, no. 5027 (1957): 1127.
124 McKendrick, 'Cinquième Revue Analytique sur les Résultats de la Vaccination Antirabique', 655.
125 McKendrick, 'Huitième Revue Analytique sur les Résultats de la Vaccination Antirabique', 11.
126 LNHO, Note sur la Rage, May 15, 1934. LON R6110, 8A/11384/11384.
127 LNHO, Note sur la Rage, May 15, 1934, 1. LON R6110, 8A/11384/11384.
128 LNHO, Note sur la Rage, May 15, 1934, 1. LON R6110, 8A/11384/11384.
129 LNHO, Note sur la Rage, May 15, 1934, 1. LON R6110, 8A/11384/11384.
130 LNHO, Note sur la Rage, May 15, 1934, 2. LON R6110, 8A/11384/11384.
131 LNHO, Twenty-Second Session of the Health Committee, October 7th–16th, 1935: Treatment of Rabies, October 5, 1935, 1–2. LON R6068, 8A/18256/11384.
132 The Fermi-types of rabies vaccine were still in use until the 1970s because they could be produced relatively easy and rapidly. See: Alan C. Jackson and William H. Wunner, *Rabies* (Amsterdam and London: Elsevier/Academic, 2007), 507.
133 LNHO, Twenty-Second Session of the Health Committee, October 7th–16th, 1935: Treatment of Rabies, October 5, 1935, 2–3. LON R6068, 8A/18256/11384.
134 LNHO, Rapport au Conseil sur les Travaux de la Vingt-Cinqième Session du Comité d'Hygiène, 26 Avril–1 Mai 1937 (Extrait du C. 219.M.159.1937.III). LON R6110, 8A/18256/11384.
135 Société des Nations, ed., 'Rapport sur l'Oeuvre de l'Organisation d'Hygiène de Juin 1938 à Avril 1939 et sur son Programme de Travail 1939 – E: Rage', *Bulletin de l'Organisation d'Hygiène* 8 (1939): 16–17.
136 Remlinger to Gautier, May 11, 1945. LON R6110, 8A/43055/11384.
137 Gautier to Remlinger, July 9, 1945. LON R6110, 8A/43055/11384.
138 About McKendrick's analytical work and research results see also: Chakrabarti, 'Living versus Dead', 418–19.

139 Gautier to Greenwood, November 9, 1945. LON R6068, 8A/43353/1167. Major Greenwood (1880–1949) worked from 1928 to 1945 as professor of epidemiology and vital statistics at the London School of Hygiene and Tropical Medicine. About Greenwood: Vern Farewell and Tony Johnson, 'Major Greenwood (1880–1949): A Biographical and Bibliographical Study', *Statistics in Medicine* 35, no. 5 (February 28, 2016): 645–70.
140 Gautier to Greenwood, November 9, 1945, 1. LON R6068, 8A/43353/1167.
141 Major Greenwood, 'Dixième Rapport sur les Données Relatives au Traitement Antirabique fournies par les Instituts Pasteur', *Bulletin de l'Organisation d'Hygiène* 12, no. 3 (1945): 319–84.
142 Greenwood, 'Dixième Rapport sur les Données Relatives au Traitement Antirabique', 373–5. About McKendrick's conclusions in 1940 see: Chakrabarti, 'Living versus Dead', 418–19.
143 World Health Organization, 'Rabies', accessed March 19, 2021, https://www.who.int/health-topics/rabies#tab=tab_1.
144 World Organisation for Animal Health (OIE), 'No More Deaths from Rabies', accessed April 1, 2021, http://www.oie.int/infographic/rabies/.
145 World Organisation for Animal Health (OIE), 'Rabies Portal', accessed April 25, 2021, http://www.oie.int/animal-health-in-the-world/rabies-portal/.
146 See also in detail: Stephan Häsler, 'Die Veterinärabkommen des Völkerbundes von 1935: Eine Vision des Veterinärraumes Europa', *Schweizer Archiv für Tierheilkunde* 160, no. 1 (2018): 37.
147 F. P. Walters, *A History of the League of Nations* (London; New York: Oxford University Press, 1952), 424.
148 Walters, *A History of the League of Nations*, 424.
149 Patricia Clavin, *Securing the World Economy: The Reinvention of the League of Nations, 1920–1946* (Oxford: Oxford University Press, 2013).
150 See Chapter 3; also: Häsler, 'Veterinärabkommen', 37–8. The Bulletin of the OIE regularly published bilateral agreements.
151 For the role of questions of agriculture and animal diseases at the conference: Häsler, 'Veterinärabkommen', 38; Ursula Reis, *Tierseuchen und internationaler Handel. Veterinärexperten im Völkerbund 1927–1935* (Universität Zürich: Lizentiatsarbeit, unpublished, 2005), 35–6; Frederick William Kohlmeyer, *The Movement toward International Cooperation in Food and Agriculture: Background of the FAO of the UN* (PhD thesis, University of Minnesota, 1954), 131–4.
152 For a contemporary overview: Louise Sommer, 'Die Vorgeschichte der Weltwirtschaftskonferenz (Genf 1927)', *Weltwirtschaftliches Archiv* 28 (1928): 340–418, for the agenda of fostering peaceful relations through economic stabilisation esp. 346–50.
153 See also: Kohlmeyer, *International Cooperation in Food and Agriculture*, 132; Amy L. S. Staples, *The Birth of Development. How the World Bank, Food and Agriculture Organization, and World Health Organization Changed the World, 1945–1965* (Kent, Ohio: Kent State University Press, 2006), 70.
154 League of Nations Economic and Financial Section: Monetary and Economic Conference, *Report and Proceedings of the World Economic Conference held at Geneva, May 4th to 23rd, 1927*, vol. 1 (Geneva: League of Nations, 1927), 181–95; Reis, *Tierseuchen*, 35–9 with analysis of veterinary-related discussions.
155 League of Nations Economic and Financial Section: Monetary and Economic Conference, *Report and Proceedings of the World Economic Conference held at Geneva, May 4th to 23rd, 1927*, vol. 2 (Geneva: League of Nations, 1927), 54. See also: Kohlmeyer, *International Cooperation in Food and Agriculture*, 134; Reis, *Tierseuchen*, 38–40.

156 Hendrikus Colijn, 'The World Economic Conference of 1927', *Annals of the American Academy of Political and Social Science* 134 (November 1, 1927): 140, 143–4. Hendrikus Colijn (1869–1944) was in office as Prime Minister of the Netherlands 1925–1926 and 1930–1939; from 1927 to 1929 he represented the Netherlands at the League of Nations.

157 League of Nations, 'Veterinary Questions, 1. General Report of the Sub-Committee of Experts on Veterinary Questions, November 4, 1930. Document C. 633.M.252.1930.II', 2, accessed April 1, 2021, https://biblio-archive.unog.ch/Dateien/CouncilMSD/C-633-M-252-1930-II_EN.pdf. About the conference and the recommendation see also: Reis, *Tierseuchen*, 40–1.

158 In his assessment of the conference from 1954, Kohlmeyer describes the League work on veterinary regulation as more or less the only lasting and effective consequence of the 1927 conference: "Reasonably satisfactory progress as a sequel to the World Economic Conference was established only in the limited area of veterinary regulations". Kohlmeyer, *International Cooperation in Food and Agriculture*, 137.

159 League of Nations, 'Veterinary Questions, 1. General Report of the Sub-Committee of Experts on Veterinary Questions, November 4, 1930. Document C. 633.M.252.1930.II', 2. See also: Reis, *Tierseuchen*, 41–2. On the convention on plant protection see: Stéphane Castonguay, 'Creating an Agricultural World Order: Regional Plant Protection Problems and International Phytopathology, 1878–1939', *Agricultural History* 84, no. 1 (January 1, 2010): 62–8.

160 League of Nations, 'Veterinary Questions, 1. General Report of the Sub-Committee of Experts on Veterinary Questions, November 4, 1930. Document C. 633.M.252.1930.II'. For a detailed discussion of the work of the subcommittee on veterinary questions: Häsler, 'Veterinärabkommen', 38–42; Reis, *Tierseuchen*, 43–87 (with analysis of the sessions). For a contemporary overview: Albert Lhoste, *Les Conventions Vétérinaires Internationales* (Paris: Vigot Frères, 1936).

161 The first subcommittee on veterinary questions consisted of Bürgi (Switzerland), Hamr (Czechoslovakia), Jensen (Denmark), Nowak (Poland), Petrovitch (Yugoslavia), Vallée (France), Wehrle (Germany), Leclainche (OIE); from the second session on additionally Kasper (Austria), Bisanti (Italy), de Figueiredo Parreiras-Horta (Brazil), Jackson (Great Britain); from the third session on additionally Mendy (Uruguay), Reid (New Zealand), and Richelet (Argentina); additional expert participants were Fischoeder (Poland), Müssemeier (Germany; spelled Meessemeier in the document). At the fourth session, new members were Voracek (Czechoslovakia) and Markowski (assistant of Nowak). League of Nations, 'Veterinary Questions, 1. General Report of the Sub-Committee of Experts on Veterinary Questions, November 4, 1930. Document C. 633.M.252.1930.II', 2–3. During the sessions of the subcommittee, the group of experts underwent several changes, see in detail about the composition of the subcommittee and the context: Reis, *Tierseuchen*, 46–61.

162 League of Nations, 'Veterinary Questions, 1. General Report of the Sub-Committee of Experts on Veterinary Questions, November 4, 1930. Document C. 633.M.252.1930.II', 3. See also: Reis, *Tierseuchen*, 61–2; Häsler, 'Veterinärabkommen', 38.

163 League of Nations, 'Veterinary Questions, 1. General Report of the Sub-Committee of Experts on Veterinary Questions, November 4, 1930. Document C. 633.M.252.1930.II', 4. For the discussions about the relations between the subcommittee and the OIE from the perspective of the latter: Office International des Epizooties, Procès-Verbaux de la Deuxième Session du Comité, Première Séance, May 15, 1928, in: *OIE Bulletin* 1 (Juillet 1927–Juin 1928), 607–11. About the relations with the OIE see also: Reis, *Tierseuchen*, 117–18.

164 Société des Nations, Comité Economique: Sous-Comité d'Experts en Matière de Mesures de Police Vétérinaire, Première Session Provisoire, Troisième Séance, January 31, 1928, 7–8. LON R2774, 10D/2049/272; about the visit of Rajchman also: Reis, *Tierseuchen*, 66–7; Cornelia Knab, 'Infectious Rats and Dangerous Cows: Transnational Perspectives on Animal Diseases in the First Half of the Twentieth Century', *Contemporary European History* 20, no. 3 (2011): 302.

165 Rajchman, in: Société des Nations, Comité Economique, Sous-Comité d'Experts en Matière de Mesures de Police Vétérinaire, Première Session Provisoire, Troisième Séance, January 31, 1928, 8. LON R2774, 10D/2049/272 (translation: CK).

166 League of Nations, Veterinary Questions. General Report of the Sub-Committee of Experts on Veterinary Questions, November 4, 1930. LON R2795, 10D/16979/1660 (Document C. 633.M.252.1930.II). In detail about the report and the process of creating the conventions: Reis, *Tierseuchen*, 84–98.

167 About the revision: Société des Nations, Comité Economique: Resumé Succinct des Travaux de la Réunion de Délégués Gouvernementaux pour l'Examen de Trois Projets de Conventions Relatives à des Questions Vétérinaires, October 15–20, 1934. LON 10A/15446/7996; Reis, *Tierseuchen*, 103–9; Häsler, 'Veterinärabkommen', 39–40. About the conference: Reis, *Tierseuchen*, 111–12.

168 United Nations Treaty Collection, 'International Convention for the Campaign against Contagious Diseases of Animals. Geneva, February 20, 1935', United Nations Treaty Collection, accessed February 19, 2021, https://treaties.un.org/Pages/LONViewDetails.aspx?src=LON&id=568&chapter=30&clang=_en (referring to: League of Nations, Treaty Series, vol. 186, 173–90). See also: Häsler, 'Veterinärabkommen', 39; Reis, *Tierseuchen*, 93–4.

169 United Nations Treaty Collection, 'International Convention Concerning the Transit of Animals, Meat and Other Products of Animal Origin. Geneva, February 20, 1935', United Nations Treaty Collection, accessed February 19, 2021, https://treaties.un.org/Pages/LONViewDetails.aspx?src=LON&id=569&chapter=30&clang=_en (referring to: League of Nations, Treaty Series, vol. 193, 37–57). See also: Häsler, 'Veterinärabkommen', 39; Reis, *Tierseuchen*, 94–5, 105–9.

170 For the context of including issues of animal protection in the convention: Reis, *Tierseuchen*, 106.

171 United Nations Treaty Collection, 'International Convention Concerning the Export and Import of Animal Products (other than Meat, Meat Preparations, Fresh Animal Products, Milk and Milk Products). Geneva, February 20, 1935', United Nations Treaty Collection, accessed February 19, 2021, https://treaties.un.org/Pages/LONViewDetails.aspx?src=LON&id=570&chapter=30&clang=_en (referring to: League of Nations, Treaty Series, vol. 193, 59–77). See also: Häsler, 'Veterinärabkommen', 40; Reis, *Tierseuchen*, 96–7.

172 For the process of signing and ratification: Reis, *Tierseuchen*, 113. See also the later evaluation of the process by the League Secretariat: League of Nations, Economic Committee: Note by the Secretariat on Veterinary Questions, May 31, 1939, esp. 2–3. LON 10A/38438/673.

173 About the aftermath of the conventions: Häsler, 'Veterinärabkommen', 40.

174 World Trade Organization, 'WTO Legal Texts—The General Agreement on Tariffs and Trade (GATT 1947)', accessed April 1, 2021, https://www.wto.org/english/docs_e/legal_e/gatt47_02_e.htm#articleXX, Art. XX; see also: Häsler, 'Veterinärabkommen', 41.

175 League of Nations, Economic Committee, Preparatory Document, Sub-Committee of Experts in the Matter of the Export and Import of Meat and Meat Preparations, Geneva, May 1935 [no day indicated], Point 1: Resolution of the London Monetary and Economic Conference, 1. LON R4379, 10A/33814/1513; about the context of the conference and the resolution see also: Reis, *Tierseuchen*, 99–101.

176 Institut International du Froid, André Lebon, Président du Comité Directeur, to the Ministère des Affaires Etrangères (France), June 8, 1928. ADLC: Unions, 429QO/498 (Bétail-Epizooties 1926–28).

177 For a contemporary study of the international trade in meat by the IIA see: International Institute of Agriculture/Institut International d'Agriculture, *Le Commerce International de la Viande* (Rome: Institut International d'Agriculture, 1936). For the significance of South American meat in international trade: Robert Greenhill, 'Shipping and the Refrigerated Meat Trade from the River Plate, 1900–1930', *International Journal of Maritime History* 5, no. 1 (1992): 65–82.

178 League of Nations, Economic Committee, Sub-Committee of Experts in the Matter of the Export and Import of Meat and Meat Preparations, Preparatory Document Drawn up by the Secretariat for the Meeting on June 25th, 1935, May 1935 (registry stamp May 23), 1. LON R4379, 10A/33814/1513.

179 About the Institut International du Froid/International Institute of Refrigeration and its activities see: LONSEA, 'International Institute of Refrigeration/Institut International du Froid', accessed March 12, 2021, http://www.lonsea.de/pub/org/257.

180 League of Nations, Economic Committee, Sub-Committee of Experts in the Matter of the Export and Import of Meat and Meat Preparations, Preparatory Document Drawn up by the Secretariat for the Meeting on June 25th, 1935, May 1935 (registry stamp May 23), 1. LON R4379, 10A/33814/1513.

181 H. [Hubertus] C. L. E. Berger (1876–1950) was professor at the University of Utrecht and head of the Dutch veterinary services; he served as president of the OIE committee from 1934 to 1939. About Berger: Universiteit Utrecht, 'Catalogus Professorum: Professoren van de Universiteit Utrecht sinds 1636: Prof. Dr. H. C. L. E. Berger', accessed May 30, 2021, http://profs.library.uu.nl/index.php/profrec/getprofdata/149/2/30/0. The other participants of the committee were: Beyro (Argentina), Boes (Belgium), Guimaraes (Brazil), Verge (France), Flückiger (Switzerland), Seoane (Uruguay). League of Nations [Economic Committee], Sub-Committee of Experts in the Matter of the Export and Import of Meat and Meat Preparations, Preparatory Document Drawn up by the Secretariat for the Meeting on June 25th, 1935, May 1935 (registry stamp May 23), 2. LON R4379, 10A/33814/1513.

182 League of Nations [Economic Committee], Sub-Committee of Experts in the Matter of the Export and Import of Meat and Meat Preparations, Preparatory Document Drawn up by the Secretariat for the Meeting on June 25th, 1935, May 1935 (registry stamp May 23), 1–2. LON R4379, 10A/33814/1513.

183 League of Nations [Economic Committee], Sub-Committee of Experts in the Matter of the Export and Import of Meat and Meat Preparations, Preparatory Document Drawn up by the Secretariat for the Meeting on June 25th, 1935, May 1935 (registry stamp May 23), 4–5. LON R4379, 10A/33814/1513.

184 League of Nations [Economic Committee], Sub-Committee of Experts in the Matter of the Export and Import of Meat and Meat Preparations, Preparatory Document Drawn up by the Secretariat for the Meeting on June 25th, 1935, May 1935 (registry stamp May 23), Annex: Import and Export of Meat and Meat Preparations. Outline of a Programme of Discussion for the Committee of Experts, 4–5. LON R4379, 10A/33814/1513.

185 League of Nations [Economic Committee], Sub-Committee of Experts in the Matter of the Export and Import of Meat and Meat Preparations, Preparatory Document Drawn up by the Secretariat for the Meeting on June 25th, 1935, May 1935 (registry stamp May 23), 3–4. LON R4379, 10A/33814/1513.

186 League of Nations, Economic Committee, Export and Import of Meat and Meat Preparations, Report to the Economic Committee by the Sub-Committee of Experts (First Session), July 3, 1935. LON R4427, 10A/15577/15577; League of Nations, Economic Committee, Export and Import of Meat and Meat Preparations, Report to the Economic Committee by the Sub-Committee of Experts (Second Session), April 30, 1938. LON R4427, 10A/33868/15577.

187 The participants of the first session in May 1935 were: Berger (Netherlands, as chairman), Bisanti (Italy), Drouin (France), Flückiger (Switzerland), Heywood (Australia), Hilton (Canada), Marczewski (Poland), Morgan (England), Piettre (IIR), Ružička (Czechoslovakia), White (USA). League of Nations, Economic Committee: Export and Import of Meat and Meat Preparations, Report to the Economic Committee by the Sub-Committee of Experts (First Session), July 3, 1935, 1. LON R4427, 10A/15577/15577.

188 League of Nations, Economic Committee, Export and Import of Meat and Meat Preparations, Report to the Economic Committee by the Sub-Committee of Experts (First Session), July 3, 1935, 2. LON R4427, 10A/15577/15577.

189 League of Nations, Economic Committee, Export and Import of Meat and Meat Preparations, Report to the Economic Committee by the Sub-Committee of Experts (First Session), July 3, 1935, Attachment: Draft Provisions for an International Sanitary Convention on the International Commerce in Meat and Meat Preparations. LON R4427, 10A/15577/15577.

190 League of Nations, Economic Committee, Export and Import of Meat and Meat Preparations, Report to the Economic Committee by the Sub-Committee of Experts (First Session), July 3, 1935, Attachment: Draft Provisions for an International Sanitary Convention on the International Commerce in Meat and Meat Preparations, 1 (Art. I). LON R4427, 10A/15577/15577.

191 League of Nations, Economic Committee, Export and import of meat and meat preparations, Report to the Economic Committee by the Sub-Committee of Experts (First Session), July 3, 1935, Attachment: Draft Provisions for an International Sanitary Convention on the International Commerce in Meat and Meat Preparations, 1–6. LON R4427, 10A/15577/15577.

192 The newly joining experts in the second session of April 1938 were: Arias (Argentina), Suarez (Uruguay), Reid (New Zealand). League of Nations, Economic Committee, Export and Import of Meat and Meat Preparations, Report to the Economic Committee by the Sub-Committee of Experts (Second Session), April 30, 1938, 1–2. LON R4427, 10A/33868/15577.

193 League of Nations, Economic Committee, Export and Import of Meat and Meat Preparations, Summary of the Discussions at the Second Session, July 13, 1938, 3–4. LON R4427, 10A/33868/15577.

194 League of Nations, Economic Committee, Export and Import of Meat and Meat Preparations, Report to the Economic Committee by the Sub-Committee of Experts (Second Session), April 30, 1938, 4–5. LON R4427, 10A/33868/15577.

195 League of Nations, Economic Committee, Note by the Secretariat, May 31, 1939, 3 (quotation marks in document). LON R4373, 10A/38438/673.

196 About the process of transition towards non-binding rules see also: Reis, *Tierseuchen*, 122.

197 League of Nations, Economic Committee, Note by the Secretariat, May 31, 1939, 3. LON R4373, 10A/38438/673.

198 League of Nations, Economic Committee, Note by the Secretariat, May 31, 1939, 3. LON R4373, 10A/38438/673.

199 The participants of the committee of experts on meat inspection were: Berger (Netherlands), Boes (Belgium), Boussard (France), Flückiger (Switzerland), Goodman (United Kingdom), Reid (New Zealand), Tinney (Australia), Viacava

(Argentina), Young (Canada); White (USA) was prevented from attending. League of Nations, Economic Committee: Methods for the Inspection of Meat. Report by the Committee of Experts to the Economic Committee, February 28, 1939, 1. LON R4460, 10A/36175/33950.

200 League of Nations, Economic Committee, Methods for the Inspection of Meat. Report by the Committee of Experts to the Economic Committee, February 28, 1939, 2. LON R4460, 10A/36175/33950.

201 League of Nations, Economic Committee, Methods for the Inspection of Meat. Report by the Committee of Experts to the Economic Committee, February 28, 1939, Annex: International Rules Relating to the Methods for the Inspection of meat Intended for the International Trade, 3–11. LON R4460, 10A/36175/33950.

202 League of Nations, Economic Committee: Methods for the Inspection of Meat. Report by the Committee of Experts to the Economic Committee, February 28, 1939, International Rules Relating to the Methods for the Inspection of Meat Intended for the International Trade, 3–11. LON R4460, 10A/36175/33950.

203 The participants of the committee of experts on the rules concerning the export and import of live animals were: Berger (Netherlands), Boussard (France), Boyle (Great Britain), Flückiger (Switzerland), Jankauskas (Lithuania), Petrovitch (Yugoslavia), Young (Canada); Manninger (Hungary) was prevented from attending. League of Nations, Economic Committee, Rules Concerning the Export and Import of Live Animals. Report Submitted to the Economic Committee by the Committee of Experts, March 25, 1939, 1, Annex: International Rules Concerning the Export and Import of Live Animals. LON R4467, 10A/37646/34956.

204 League of Nations, Economic Committee, Rules Concerning the Export and Import of Live Animals. Report Submitted to the Economic Committee by the Committee of Experts, March 25, 1939, Annex: International Rules Concerning the Export and Import of Live Animals, 3. LON R4467, 10A/37646/34956.

205 League of Nations, Economic Committee, Rules Concerning the Export and Import of Live Animals. Report Submitted to the Economic Committee by the Committee of Experts, March 25, 1939, International Rules Concerning the Export and Import of Live Animals, 5–6. LON R4467, 10A/37646/34956.

206 League of Nations, Economic Committee, Rules Concerning the Export and Import of Live Animals. Report Submitted to the Economic Committee by the Committee of Experts, March 25, 1939, International Rules Concerning the Export and Import of Live Animals, 6–8. LON R4467, 10A/37646/34956.

207 League of Nations, Economic Committee, Methods of Preserving Products of Animal Origin, Note by the Secretariat, June 9, 1939. LON R4468, 10A/37431/37431.

208 League of Nations, Economic Committee, Methods of Preserving Products of Animal Origin, Note by the Secretariat, June 9, 1939, 2–3. LON R4468, 10A/37431/37431.

209 League of Nations, Economic Committee, Methods of Preserving Products of Animal Origin, Note by the Secretariat, June 9, 1939, plus summary 3–4; Annex 1: "Scheme for a Study of the Methods of Preserving Meat and Meat Products…". LON R4468, 10A/37431/37431.

6 Veterinary internationalism in times of crisis

The Office International des Epizootics, 1927–1947

In the late 1920s and 1930s, a number of international projects and institutions dealt with different aspects of cross-border animal disease problems. Compared to the pre-war period, the debates on the subject had created a new dimension of international awareness, initiatives, efforts of cooperation and networking activities. After a break of 16 years, the series of international veterinary congresses was taken up again, with the Eleventh International Veterinary Congress taking place in London in 1930. Two congresses followed, in 1934 in New York and in 1938 in Zurich. In 1931, the series of International Congresses of Comparative Pathology resumed after the first meeting in 1912; further congresses followed in 1936 in Athens and in 1939 in Rome. The location of the following International Veterinary Congress of 1942 should be Berlin.[1] The international congresses, platforms for the growing international scientific community, showed how the academic discipline of veterinary medicine developed and diversified in its interdisciplinary links with medical, scientific and technical academic fields. International institutions, such as the League of Nations and its cooperating organisations with their expert committees, included the issue of transboundary animal diseases on their agenda to negotiate international agreements and discuss effective approaches. The most important institution in the field of veterinary internationalism during the late 1920s and 1930s was the OIE in Paris. The new international office quickly established itself as a hub in the organisation of worldwide information transfer, in the networking of various institutions and in international professional exchange on the topic of animal diseases.[2]

Despite the political and economic tensions of the 1930s, the OIE built up and continuously expanded a system of regular collection and transmission of information on animal diseases with a thematic and geographical scope that had not previously existed at the international level. Given its limited size and financial means, the new international office was from the beginning to a large degree dependent on the commitment, initiatives, personal networks and resources of its staff and associated experts. Beyond its role in organising and distributing disease information, the statutes allowed the

DOI: 10.4324/9781003033998-6

office to act as a promoter and coordinator of research activities in the field of infectious diseases of livestock and their prophylaxis.[3] When from 1937 on outbreaks of foot-and-mouth disease affected most countries in Europe, the experts at the OIE started to conceptualise a new effort of transboundary collaboration against the spread of the epizootic; the project, however, was not realised any further. During the Second World War, the OIE staff tried to continue its work as much as possible to ensure the institutional continuation of the office. New international organisations were founded during the war and in the post-war period to tackle the problem area of animal diseases in a global perspective. The developments resulted in questioning the inherent Eurocentrism of pre-war veterinary internationalism, and therefore brought about new horizons for international cooperation to approach animal diseases in their global dimension.

Building structures and networks of veterinary internationalism in times of crisis

The OIE began its activities in March 1927 with the first meeting of its international committee with 24 countries represented.[4] Henri de Roo, the Belgian veterinary inspector-general, was elected as president of the OIE committee; Emmanuel Leclainche, head of the French veterinary services at the Ministry of Agriculture and the mainorganiser of the new office, as OIE director.[5] In its first years, the OIE registered an increase in the number of member states. This seemed to indicate a certain acceptance of the new international institution. But the changes in the list of the member states, published by the OIE in its regular Bulletin next to the front page, also revealed the office's difficulties. For the operation of the office, it was important to involve as many member states as possible, because of the dependence on member contributions and to ensure international legitimacy. The office also needed an extensive network of communication with national veterinary services and research institutions to effectively organise information on the global animal disease situation. The first annual volume of its Bulletin listed 44 member states, including several mandates and colonies; the majority were European.[6] Large parts of the world did not take part; North American countries (Canada and the U.S.) did not participate, nor did China. The Soviet Union, however, was listed as a member.[7] In the following years, several governments, including those of mandates and colonies, newly joined.[8] The number of member states oscillated during the early 1930s: some smaller states as well as colonial and mandate administrations started to rethink their membership in international organisations such as the OIE for financial reasons. The governments of the Baltic States sent a joint request to the OIE secretariat asking to be represented by one expert for all three countries; Leclainche rejected for financial reasons.[9] Participation in technical-scientific international projects such as animal disease control was apparently not considered indispensable by some governments. In 1932/33, the effects of the worldwide economic crisis on the

OIE became also apparent in discussions about the financial contributions of member states. The problem was connected with the status of colonies, mandates and protectorates in international organisations such as the OIE. The British Foreign Office was reluctant to pay the contributions to the OIE on a gold basis for Britain and its colonies and mandate territories.[10] Leclainche was increasingly concerned about the debates and insisted on the necessity of contributions in gold francs. In February 1933, in a report to the French Ministry of Foreign Affairs, he pointed out that many of the British colonies and protectorates had actually so far never paid member fees, in spite of being listed as members of the OIE. Leclainche stated that the OIE had not even protested officially, since it aimed for good relations in veterinary cooperation. The ongoing economic crises, Leclainche worried, seriously endangered the OIE's existence and international veterinary work in general.[11] Indeed, several British colonies and protectorates left the OIE in 1932/33, giving financial reasons, among them some with frequent problems of epizootics, such as British Somaliland.[12] For many British colonial veterinary administrations, engaging in international projects seemed not to be an issue of high priority. In the late 1920s, the British Ministry of Agriculture had built up its own network of Imperial Agricultural Bureaux, including the Imperial Bureau of Animal Health (since 1938 the Commonwealth Bureau of Animal Health). The Bureau was associated with the Ministry's Veterinary Research Laboratories in Weybridge, and its aim was to collect and exchange information on veterinary research and coordinate imperial research activities.[13] Still, even though some member states left the OIE, the situation of the office in the mid- and late 1930s turned out not quite as dire as Leclainche had feared. The OIE gained a number of new members during the mid-1930s, among them countries with an important livestock sector, such as Uruguay and the Union of South Africa. Other member states left when they ceased to exist as independent states, such as Austria in 1938. In 1939, the list of members again consisted of 44 states.[14]

In its first decade, the OIE operated by following the tasks of an international office as it had been outlined in its founding documents. The OIE's international committee, its decision-making body, met annually in Paris for the general session to discuss the office's institutional policies and to issue resolutions on veterinary-related topics. The international committee consisted of the permanent delegates as representatives of the member states and was chaired by the acting committee president: from 1927–1929 Henri de Roo (Belgium), 1929–1933 Ferenc Hutyra (Hungary), 1933–1934 Erwin Wehrle (Germany), 1934–1939 H. C. L. E. Berger (Netherlands), and from 1939–1949 Gottlieb Flückiger (Switzerland). Most committee members were the heads of the respective national or colonial veterinary services or national research institutions. In his annual report for each session, OIE director Leclainche outlined the institutional situation of the office, its finances, the main lines of activity, the progress of its publications, and its liaison efforts with national institutions and international organisations. Usually, Leclainche also provided a report on the worldwide situation of epizootics, and several of the attending experts gave introductions into specialised topics for the discussion. The committee discussed and published technical-scientific resolutions on diverse animal disease problems, on disease prevention and on veterinary legislation. The aim was, as Gottlieb Flückiger pointed out, to provide constant and non-binding expert recommendations to the—often hesitant—governments, and thus gradually but sustainably improve the worldwide situation concerning animal diseases.[15] The OIE's permanent bureau in Paris consisted only of a few members of staff. In October 1938, the French Ministry of Foreign Affairs established an evacuation plan for all international offices in case of war, and in this context also contacted the OIE with questions about the scope of the office. Leclainche answered that the OIE permanent bureau consisted of himself, two secretaries and two text editors.[16] They organised the office's worldwide correspondence and the meetings of the international committee, and they were responsible for editing the Bulletin and the animal disease statistics. In 1934, on the tenth anniversary of its founding convention, the committee president Berger, who had followed Wehrle in this position, summarised the past 10 years as a period in which the office steadily became accepted as a significant institution on a global scale. He highlighted the essential role of its director: "L'Office, c'est Leclainche".[17] Indeed, the OIE's work in the first two decades of its existence was shaped by the ideas, the professional commitment and the veterinary conceptualisations of Leclainche, whose opinions and directions were rarely questioned in committee meetings. Even more so, despite the fears for the OIE's existence, its situation stabilised during the 1930s. In 1936, the committee president Berger could announce that the annual contributions of those countries which paid on a gold basis could even be reduced by 10 percent.[18] At the beginning of 1939, the OIE bought a building to accommodate the office in the Rue de Prony, near the Parc de Monceau, which is still the headquarters of the OIE today.[19]

The stabilisation of the OIE opened opportunities for activities in the late 1930s, even though for a small international institution with limited staff and financial means to put meaningful anti-disease concepts into practice on a broad level presented a challenge. The OIE statutes prohibited any direct intervention into the sovereignty of a member state.[20] Still, the OIE aims and working programme outlined in Article 4 was ambitious:

a. To promote and co-ordinate all experimental and other research work concerning the pathology or prophylaxis of contagious diseases of livestock for which international collaboration is deemed desirable;
b. To collect and bring to the attention of the Governments or their sanitary services, all facts and documents of general interest concerning the spread of epizootic diseases and the means used to control them;
c. To examine international draft agreements regarding animal sanitary measures and to provide signatory Governments with the means of supervising their enforcement.[21]

Given these possible working areas, the OIE faced several questions: how to initiate activities without intervening into the national veterinary policies, how to address the cross-border dimension of epizootics in conceptual and organisational terms, how to secure further funding for projects, and how to deal with with other international organisations and initiatives, such as the parallel efforts of the League of Nations on trade-related veterinary matters. In the OIE general session of May 1931, the OIE committee vice-president Moritz Bürgi, head of the Swiss Federal Veterinary Office, warned of high expectations:

> The questions which we have provoked are eminently complex ones, and they demand for very exceptionally complete and definitive solutions. Almost always, we achieve only partial results, and therefore our activities must be methodically extended over time and ceaselessly adapted to newly emerging conditions.[22]

Leclainche and his staff focused, above all, on the tasks outlined in Article 4b of the statutes, to collect, organise, and further distribute information concerning the situation of animal diseases—the traditional core tasks of an international office, and significant for governments to take action to prevent the cross-border spread of epizootics. The OIE established from scratch a regular information transfer through its monthly Bulletin, information that was researched, organised and edited by its staff, a process for which the office needed a close communication network of governmental administrations, research institutions, international organisations, and individual researchers. The main addressees were veterinary, scientific and public health professionals in governmental positions, as well as research institutions. Information of the OIE on animal diseases constituted, in this context, the news concerning the office, original research articles written for the Bulletin, official reports on the activities of veterinary services and international institutions, legal regulations, veterinary notifications, reports on international meetings and conferences, and reviews and extracts on diverse animal disease issues, as well as statistics on the animal disease situation. Further, the Bulletin reported on all activities of the OIE and especially provided the documents concerning the proceedings of the annual OIE committee sessions (reports, documents, minutes, and resolutions of the meetings). From May 1935 on, the proceedings of the sessions were published in separate volumes together with the preparatory documents. The aim mentioned in Article 4c, to advance the work on international conventions on veterinary sanitary matters had—for the moment—become the task of the League of Nations' expert committees. The core of the OIE's work was therefore to ensure a regular organisation and transfer of information, and at the same time to act as a platform for regular exchange between veterinary administrations. The committee produced recommendations on a number of specific and current disease issues in their annual meetings. The aim was to provide a regular, reliable, systematic and transparent overview on animal diseases and their handling based on the data of the national sanitary services of the participating countries. This not only forced the member states to report regularly on the respective situation of animal diseases. It also made information on any form of national veterinary 'progress' or its absence internationally observable, as well as comparable with other national veterinary sanitary services. The task of organising a regular information transfer on the animal disease situation therefore presented the OIE with considerable challenges. In a contribution to the first volume of the Bulletin, Leclainche emphasised that the inadequacy of national documentation of diseases in export countries was frequently used by other governments as an argument to justify import restrictions.[23] Consequentially, the OIE started already from the beginning to work on standardised templates for the organisation of information, including minimum and mandatory requirements of data structures on a number of diseases, such as rabies, glanders and foot-and-mouth disease.[24] At the annual general meetings of the OIE committee, Leclainche regularly reported on the development of this project. The aim was to cumulate and evaluate national data as broadly and precisely as possible. For the OIE staff, however, the project of setting up such an international overview and providing a reliable and regularly updated picture of the global animal disease situation proved extremely difficult. At the OIE committee meeting in May 1931, Leclainche pointed out that many national disease reports arrived irregularly at the OIE; furthermore, often, the incoming disease statistics were incomplete.[25] One year later, in May 1932, he explained that, in order to be of realistic documentary value, the incoming national reports and statistical data had to comply with two conditions: they had to clearly reflect the national sanitary situation, and they had to be comparable from country to country.[26] Otherwise, as he outlined again in the session of 1933, the movements of animal diseases beyond national borders and the appearance of new disease sources could not be adequately tracked.[27]

Similar to the international veterinary congresses of the 19th century, the problem of non-existing common international standards for assessing disease information was a frequently discussed topic at OIE committee meetings. In May 1931, Leclainche suggested integrating a new series of graphics and maps with the aim to visualise the global situation of epizootics. The graphics should illustrate the global movements of epizootics and at the same time demonstrate the effectiveness of the respective veterinary measures.[28] A year later, in 1932, Leclainche outlined that it had been impossible for the OIE to acquire sufficient and usable data for producing disease maps: there could never be more than three countries described graphically at once. Also the question of internationally uniform sanitary bulletins and of informative animal disease statistics on the international level was difficult to be advanced.[29] At the OIE committee meeting of May 1933, Leclainche showed himself as becoming increasingly worried at the incoherence and incompleteness of incoming national data material, which formed the basis of the production of OIE information. He pointed out that the OIE met with considerable resistance, lack of knowledge, and incompetence on the part of many national veterinary services, whose staff often ignored the suggested scheme of international collaboration. This insufficient collection of data constituted a dangerous problem, Leclainche argued, especially if a serious epizootic were to occur on a transboundary scale—it would be impossible to track down the origin, movement and localisation of a disease. The OIE staff, he pointed out, continued with its core task of publishing the incoming data as best as possible, but it was difficult to present the comprehensive picture of the animal disease situation on a global scale that they had hoped to provide.[30] This assessment was, to some extent, a setback with regard to the central task of the office, namely to set up a highly efficient, transparent, comprehensive and globally oriented information transfer to improve animal disease control in general and prevent the cross-border spread of diseases. Furthermore, the aim of the OIE was certainly to provide information on a broad, if possible global scale. The global nature of the problem area and the shortcomings in addressing this globality in veterinary work were clearly recognized. In a contribution to the Bulletin volume 1932/33, Leclainche noted a clear change since the late 19th century in the emergence and spread of animal diseases on a global level. He argued that, apart from foot-and-mouth disease, the 'old' diseases were on the decline or at least well-researched in many parts of the world, even though Asia continued to be a problem in this respect.[31] He warned against other, previously rather marginal dangers from animal diseases that were increasingly manifesting themselves globally, and against other "diseases of the future".[32] These included also zoonotic diseases. As Leclainche outlined, conventional veterinary sanitary anti-disease methods were not sufficient anymore to address these future dangers.[33] The topics of the articles and research reports included in the OIE Bulletin increasingly also reflected the global dimension of animal disease control. The discussions at the OIE committee, however, often focused on disease problems and containment strategies that were mainly of European relevance.[34] The tension between the European-dominated orientation and organisation of international work and the global dimension of the animal disease problem with its possible consequences thus was also present in the work of the first permanent institution of veterinary internationalism.

Projects and networks of international veterinary cooperation in the 1930s

In the mid- and later 1930s, the OIE committee and its director Leclainche increasingly began to explore the scope for promoting international research activities, in accordance with article 4a of the OIE statutes. There were several options for pursuing activities—to form closer networks of cooperation with national or other international institutions, to financially support research projects, or to act as the coordinator of joint international initiatives. Leclainche was keen to emphasise the OIE's core technical-scientific agenda, to underline its autonomy and political neutrality and to differentiate the office from other international bodies. This aim became clear, for instance, on occasion of a request for cooperation from the part of the IIA. Recent research has underlined the links between the IIA and fascist influences on agrarian internationalism.[35] The OIE and IIA had occasionally cooperated in the early years, with the OIE giving support in organising animal disease information on a questionnaire of the IIA concerning diseases in tropical countries.[36] Carlo Bisanti, the Italian delegate at the OIE committee and head of the veterinary services of the Italian Ministry of the Interior, also acted as the OIE representative at the IIA general conferences.[37] IIA president Giuseppe de Michelis approached the OIE in spring 1933, during the preparations for the London Monetary and Economic Conference, asking about possible cooperation with regard to the topic area of commerce in agricultural products. Leclainche declined, pointing out instead to the results of the work of the League subcommittee on veterinary questions. In the committee meeting of May 1933 he outlined that the OIE operated with a strictly scientific profile and on a politically neutral basis:

> I replied to Mr de Michelis that our Office deals with international exchanges only from the point of view of their sanitary consequences, and that it is in no way qualified to address economic questions.... I believe I have interpreted your wishes to mean that you wish to maintain active relations as to maintain active relations with the various international organisms, but do not want to set aside the programme that is assigned to you by your statutes.[38]

As a consequence, the OIE did not cooperate with the IIA in the conference preparations. Leclainche's reasoning separated scientific and economic questions, even though the economic relevance of animal diseases was the main reason for the office's existence. The effort to create an image as a scientific and politically neutral institution dealing with technical-scientific questions of effective veterinary measures and with factual information transfer was obviously of great importance to Leclainche, as it served the cultivation of relations with all sides. The OIE was in contact or maintained cooperative activities with further international organisations. As mentioned before, there were links between the OIE and the League of Nations' expert committees for veterinary matters.[39] In several points, the thematic focus of the OIE and the League subcommittees overlapped, especially in case of the first subcommittee of 1928–1931. The first international convention produced by the subcommitee, the International Convention for the Campaign against Animal Diseases, incorporated with regard to the topic of information transfer the OIE recommendations on the desired content and organisation of national veterinary bulletins.[40]

After 1935, the OIE was no longer represented directly in the League committees, but many participants in the committees were also national delegates in the OIE committee. In the second meeting of the OIE committee in May 1928, the OIE delegates discussed drawing a clear line between scientific questions and economic or political considerations in international projects. They agreed on the principle that the objectives of the OIE should be primarily scientific: while the meetings organised by the League subcommittee concerned an international convention project to improve commercial relations, the OIE, in contrast, was an organisation with long-term aims of cooperative work on a multiplicity of scientific and technical questions.[41]

Another field of possible institutional cooperation was the LNHO project for standardising biological substances. The LNHO Permanent Commission on Biological Standardisation met biannually from 1924, chaired by Thorvald Madsen, the president of the League Health Committee. Its work continued even during the 1940s and was taken over by the WHO.[42] In close cooperation with researchers from the laboratories of the National Institute of Medical Research in London and the Statens Seruminstitut in Copenhagen, the commission defined standards for biological substances.[43] The question of cooperation between the OIE and the permanent commission came up repeatedly in the general sessions of the OIE committee. In June 1931, Leclainche was invited to participate in a meeting of the permanent commission but declined because, as he stated, he did not have an official mandate from the OIE committee.[44] In a note to the permanent commission, Leclainche declared the OIE's support for their work, on which the OIE had collected documentation since 1928, and summarised the challenges of the issue from the perspective of veterinary medicine. Especially in the cases of rabies, brucellosis, hog cholera and anthrax, the many different diagnostics and vaccinations created, in his view, immense problems. Leclainche stated that the existing incoherence on so many levels was dangerous, since it not only risked the lives of animals and humans but potentially also contributed to the further spread of diseases.[45] In spring 1932, Madsen contacted Leclainche again about a possible cooperation of the OIE with the LNHO project,[46] and the OIE committee nominated him as representative on the permanent commission.[47] Two years later, at the OIE general session in May 1934, Leclainche reported that he considered the work of the permanent commission to be too slow, and veterinary medicine needed quicker results for its international work.[48] The OIE Committee therefore issued a resolution to initiate its own investigations on the conditions for international cooperation in the standardisation of biological products with suitable laboratories.[49] In particular, Leclainche outlined that there was the urgent need for a standard of tuberculin for both veterinary and human medicine. To develop an international tuberculin standard for animals promised more consistent diagnostic results and increasing certainty about the worldwide situation of bovine tuberculosis.[50] However, in

1928 the League's permanent commission on standardisation had already set up a standard for tuberculin. When the OIE decided to begin its own examination of the matter, it started to seem, as Leclainche outlined, that the OIE counteracted the work of the League.[51] Leclainche wrote an explanation to Raymond Gautier of the League Health Section, who would in 1939 become Rajchman's successor as acting medical director. Veterinary medicine, Leclainche argued, urgently needed practical solutions to diagnose tuberculosis in animals, pointing out that the 1928 standard was primarily employed for the diagnosis of the disease in children.[52] Finally, Madsen and Rajchman promised their support to the OIE project and their intention to invite the OIE to cooperate with the League's permanent commission on standardisation the next time tuberculin was on the agenda.[53] The OIE started an international inquiry among specialised research institutions into the question of tuberculin preparation. The results only confirmed the complexity of the question. It turned out that the development of a standard and its international implementation would indeed require a long process of research, since private laboratories produced and supplied a multiplicity of different types of tuberculin.[54] In 1937 the OIE committee decided to drop the question of an international tuberculin standard. It suggested, instead, to transfer the question of a standard diagnostic tool for brucellosis to the Royal Veterinary College in London.[55] The initiative showed that the OIE's limited resources and organisational structure as an international office made it difficult for its staff to be part of research initiatives on internationally relevant veterinary topics, despite the close networks with other institutions and within the international scientific community.

An "international entente" to address foot-and-mouth disease?

An opportunity for a new international project—and for a new role of the OIE—came in 1937 with the outbreak of a severe foot-and-mouth disease epizootic in Europe. For several years, foot-and-mouth disease had not been a major problem for most European countries. The new massive appearance of foot-and-mouth disease manifested first in Southern France in April 1937, introduced by animal importations from Northern Africa. Serious outbreaks followed in Switzerland, Belgium, Luxembourg, the Netherlands, Denmark, Germany, Poland and Czechoslovakia. The disease traveled eastwards to Russia and the Balkans.[56] Most European governments took rigorous control measures against the disease on national levels, but even with strict veterinary sanitary measures and border controls in force, the virus spread rapidly beyond national boundaries. The outbreak also showed again the significance of imperial infrastructures and trading networks in the distribution of epizootics. For the OIE, it provided an opportunity to demonstrate its relevance for the member states in an acute epizootic crisis. The office followed the news about the disease situation intensely and published incoming information, including national reports and developments in research. The office's telegraphic services began to regularly notify

European countries about the development of disease outbreaks and their spread.[57] However, in the course of 1937, the spread of the disease did not stop. The OIE convened an internal meeting on the situation in early 1938, with Leclainche as OIE director, Berger as committee president, Bisanti as vice-president, and Flückiger, the successor of Bürgi as the head of the Swiss federal veterinary administration, as treasurer.[58] They decided, according to Article 4a of the OIE statutes, to suggest a new international initiative in cooperation with the French National Veterinary School at Alfort. The aim was to establish an international control programme dealing with the methods of active immunisation for disease prophylaxis currently in use in different countries.[59] In February 1938, the OIE invited the member delegates with a circular letter to participate in the project. The letter explained that one main motivation for the initiative was indeed the ongoing spread of the disease—countries with insufficient veterinary sanitary systems and without access to vaccines, it outlined, endangered the livestock of all countries, especially of neighbouring ones. The project proposed the organisation of experimental testing of the active immunisation methods used in different countries to find a viable method of protection. The French government would cooperate by providing the research facilities at the Alfort laboratories; the OIE would carry the expenses for the personnel and the animals to be used.[60] With a new international project, the OIE intended to take up an active role in supporting and accelerating the process of finding a suitable, effective tool of immunisation which allowed a usage in many countries.

At the same time, research concerning foot-and-mouth disease vaccines was making progress, especially in Germany. Since its foundation in 1910, the world's oldest research institute for virology, the Staatliche Forschungsanstalten on the island of Riems at the coast of the Baltic sea, focused in particular on foot-and-mouth disease research.[61] The National Socialist government promoted research on foot-and-mouth disease in Riems because of worries that the pathogen might be used in biological warfare against Germany.[62] The experiments on foot-and-mouth disease vaccines during the 1930s were conducted by a group of scientists led by Otto Waldmann, the director of the Riems institute.[63] Waldmann was since 1920 the successor of Friedrich Löffler at Riems and enlarged its laboratories and research activities. During the Second World War, the Riems institute would begin with the development of a foot-and-mouth disease virus type as a biological weapon, which would not be used.[64] In 1938, the laboratories at Riems successfully produced a new foot-and-mouth disease vaccine type based on inactivation through formaldehyde and heat adsorbed on aluminiumhydroxide, which promised a broad spectrum of immunisation.[65] The vaccine was first used experimentally in the winter of 1938 in Germany. Subsequently, experiments with the vaccine started in Switzerland, where even rigorous border controls and immediate comprehensive efforts of culling had failed to prevent the repeated emergence of the disease.[66] Still during 1938, the epizootic continued to spread across borders and with considerable speed throughout Europe. The XVIII International

Congress of Agriculture in June 1939 took up the discussions on foot-and-mouth disease and immunisation methods.[67] The location of the event was Dresden—similar to the Italian government, the National Socialist government cultivated an active policy of hosting international events as opportunities for propaganda and networking, especially during the late 1930s.[68] Foot-and-mouth disease was one of the prominent topics. Both Leclainche and Flückiger highlighted in their conference contributions the idea of finding forms of international cooperation coordinated by the OIE to ensure a broad usage of the new Waldmann vaccine.[69] In its resolutions, the congress noted that it was indeed advisable to advance common actions and arrangements of all countries interested in fighting foot-and-mouth disease.[70]

The OIE started to conceptualise a new international project on vaccine production and distribution. The plans concentrated on the question how to coordinate the transboundary usage of the Waldmann vaccine type in future within a narrow time frame. The OIE committee session in May 1939 voted for the resolution that "it is indispensable that an international entente is realised with the aim to ensure in practice the preparation and the utilisation of the vaccine".[71] In late August 1939, Flückiger, the new OIE committee president, and Leclainche as director addressed a circular letter to all OIE delegates. It suggested to address foot-and-mouth disease on the international level with a concerted transboundary project of vaccine production and distribution.[72] The problem, as Leclainche had already outlined at the Dresden congress, was twofold: firstly, the vaccine was expensive in production (since it required specialised research infrastructure and huge numbers of cattle), and secondly, in case of an outbreak, it needed to be applied very quickly and comprehensively, being only preservable for a short period. Outbreaks could suddenly appear anywhere at any time, but sometimes the disease did not emerge for several years.[73] Because the numbers of livestock were increasing in many countries, a comprehensive transboundary vaccination campaign would have to be organised to have the vaccine available in short time periods. If every country had a chance to get the necessary amounts of vaccines through its cooperation partners, Leclainche and Flückiger argued, there was a far better chance than before to gain control over primary outbreaks and seats of the disease.[74] Research institutes producing the vaccine for national usage existed, according to the OIE, in Italy, Germany and Denmark.[75] Both the Danish and the German government had confirmed their readiness to cooperate and also to provide vaccines to other countries.[76] According to the proposal, a country affected by an outbreak of foot-and-mouth disease should get help from abroad, a process that should be settled through an international arrangement—to centralise the vaccine production or increase the number of production institutes.[77] The circular included a list of questions to the OIE member governments to assess the possibilities and requirements for international cooperation addressing foot-and-mouth disease through a coordinated vaccination campaign. The questions asked were about the member governments' opinions and possible objections concerning the cooperation project. The OIE staff planned, after evaluation of the responses, to organise an international conference on foot-and-mouth disease in autumn 1939 to discuss and decide on how to proceed, and whether an international agreement on the issue could succeed.[78]

Because of the timing of August 1939 it was quickly clear that the initiative of a new transboundary veterinary cooperation would have few chances to be realised, with the German National Socialist government supposedly in a rather central position. Though several neutral governments communicated their approval, the plan was dropped when the war began in Europe several days later.[79] Foot-and-mouth disease continued to be a problem throughout the war years.[80] Still, the project of 1939 showed the emergence of a possible new and more active role for the OIE as an international coordinator of a project of vaccine production, distribution and surveillance. The suggested vaccination project combined the idea of centralised international coordination with direct national and local implementation, aiming to provide an effective and fast transboundary coverage of vaccination. Its realisation would therefore have required some level of transparency in the collaboration between the respective national scientific and veterinary systems, as well as the willingness for building up regular lines of cooperation and communication. Indeed, it would have been a project of 'organising trust' on a strategic and practical level, since it intended to manage interdependencies in vaccine production and distribution across borders, not to mention the question of funding. The political dimension in the planned international project, especially concerning a partial dependency on the German National Socialist government because of the production of the Riems vaccine, was not at the fore of discussions at this early point of the project, at least not in official written form. OIE director Leclainche frequently underlined during the meetings in the 1930s the non-political and neutral character of the office; this certainly helped to keep up a minimum of international communication in so-called technical fields such as the control of epizootics. Several governments showed interest in international cooperation concerning foot-and-mouth disease vaccination in spite of the ongoing severe political tensions and ideological gulfs in the international system. Therefore, during the late 1930s, the networking activities of the OIE and the office's efforts in addressing the transboundary dimension of foot-and-mouth disease opened up new potential roles for the OIE as the initiator and coordinator of international projects. Collecting and organising the international transfer of information, the traditional and central task of the office, was not a sufficient response to an acute crisis of epizootics of transboundary relevance, where more comprehensive and transboundary responses were needed.

The Second World War and veterinary internationalism

The war had a profound impact on international organisations such as the OIE in terms of institutional existence and work. Many institutional and professional structures as well as networks of international organisation and cooperation were shut down. The control of animal diseases was relevant for the war economy of all countries, since veterinary services were needed to maintain the flow of food supplies and transport animals, which were still used in large numbers despite the advanced motorisation of armies. This brought the OIE into the focus of the National Socialist government and its plans for German-dominated international organisations. When the German army occupied Paris in June 1940, the situation for the staff members of the OIE, as for those of other international offices and organisations, became difficult. Leclainche was staying in Toulouse at the time, and the OIE in the hands of the general secretary Georges Le Louet.[81] What followed was a struggle for the institutional survival and independence of an international organisation in times of war under German occupation. The German National Socialist government and the military authorities, both in Berlin and in Paris, contemplated during the whole period of occupation how to use the OIE's infrastructure for their purposes. In late 1940, the Vichy Ministry of Foreign Affairs inquired in a letter to Leclainche in Toulouse whether the OIE actually still existed; the Irish government had asked how to submit its annual contribution.[82] Leclainche underlined in his answer vehemently the necessity of doing everything possible for the continuing existence of the OIE. On this occasion, he gave an overview of the proceedings of the past months and the situation of the international office under German occupation from his point of view.[83] He outlined that the future of the OIE seemed doubtful, given that most lines of communication to the world outside France had been restricted, and printing the OIE publications had become difficult. For the moment, organising meetings of the OIE committee as the decision-making body was not possible. As Leclainche explained, after the occupation of Paris the German military authorities had closed and sealed all premises of international offices.[84] In October 1940, Leclainche had visited Paris and contacted the German authorities in order to secure the continuation of the OIE.[85] In his answer to the Vichy government, he provided a list of arguments in favour of the continuation of the office, including the ability to resume its activities at once since it had acquired the necessary infrastructure and its own building.[86] The regular funding of the office, as he outlined, constituted a problem. Many member states did not pay their contributions anymore, and there was confusion about the way of payment.[87] The Vichy government itself finally decided in mid-1942 to continue its payments for the office.[88] Both Flückiger and Leclainche tried to establish working relations with the German authorities in order to be able to resume the activities of the OIE and secure its existence, as Leclainche wrote to the Vichy Ministry of Foreign Affairs in early 1941:

> ...on the other hand, I had very friendly personal relations with Doctor Friedrich Weber, Director in the Ministry of the Interior in Berlin and

> personal friend of Chancellor Hitler. I know that Dr. Weber is very well disposed towards the Office and that he wishes it to take up its activities as soon as possible. I have seen the echo of these dispositions during my visits to the German authorities in Paris, and I believe I have understood that one is thinking about keeping the Office in Paris.[89]

Friedrich Weber, whom Leclainche mentioned, had already been part of the Munich Putsch in 1923 and had made his career in the National Socialist administration and the SS. Since 1934 he was "Reichstierärzteführer" and worked in the Ministry of the Interior in Berlin.[90] In order to keep up the institution and the work of the OIE in Paris, Leclainche did not have many alternatives other than to take up contact with the German occupying authorities. The Foreign Office in Berlin confirmed to the German Embassy in Paris that it had no objections concerning Leclainche personally: he had, as they outlined, as head of the French veterinary services in 1919/20 cooperated well with German veterinarians in negotiations on the transfer of livestock in context of the Paris Peace Conferences.[91]

Eventually the Swiss president of the OIE committee, Flückiger, negotiated that the OIE was able to again resume its work to a limited extent. Flückiger travelled to Berlin for talks and, as a result, on February 27, 1941 a meeting in Paris decided on the OIE's future.[92] As Leclainche wrote to the Ministry of Foreign Affairs in Vichy in late 1941, he himself could not attend this meeting because of a dispute between the Vichy and the German authorities over his passport papers. Therefore, Flückiger as committee president and Le Louet as general secretary of the OIE took part, together with Camille Boussard as the representative of the Vichy Ministry of Agriculture, and came to an agreement with Weber and two representatives of the German military command.[93] Weber confirmed the continuing existence of the OIE in Paris and promised its operation under German control: all major staff decisions had to be approved, the language of correspondence had to be German, all reports and work needed to be coordinated with the German authorities, and a German veterinarian would be present in the building as "liaison and control agent".[94] The OIE was thus able to resume its work in a controlled, limited and German-dominated form. The seals were removed from the doors and the keys of the building handed over to Flückiger.[95] Leclainche returned to Paris in June 1941, later than planned because of delays in getting adequate papers for travelling. Through Flückiger he had learned that the German authorities, and Weber in particular, were beginning to lose patience because of his absence.[96] In late 1941, Leclainche confirmed to the Vichy Ministry of Foreign Affairs (at the latter's request) that it did not seem to him that a transfer of the OIE to somewhere outside France, to Berlin for example, was intended.[97] In addition, 15 member governments had paid their annual fees for 1940, and three even for 1941. No member had so far officially declared its withdrawal.[98] But the relations between Leclainche, the Vichy government and the German authorities remained tense. In early 1942, the Vichy Ministry of Foreign Affairs criticised in an internal paper the arrangements with the German authorities, especially the German presence and control of the OIE as well as the correspondence in the German language, and stated that the independence of the office was impaired and these changes were in contradicton with its statutes.[99] Since November 1943, the Vichy Ministry of Foreign Affairs demanded comprehensive and detailed information about the OIE staff, their tax payments, the building, and the finances of the OIE. Leclainche explained that such an inquiry constituted an intrusion into the independence of the office according to its statutes.[100] The question of an international office's status and independence had obviously become a tool for exerting a certain amount of pressure. The German authorities, likewise, were in the long run interested in more than controlling and infiltrating the OIE and its work. From 1941 on they tried to convene a meeting of several remaining members of the OIE committee, which should assemble in 1942 to bring together delegates from German-controlled regions as well as from neutral countries. Leclainche explained after the war that the OIE successfully managed to avert such a meeting.[101]

The administrative correspondence between the German Foreign Office and the German Ministry of the Interior in Berlin showed that the existence of the OIE was far from safe during those years. From late 1941 onward, there were debates how to revive the international exchange of information about epizootics in order to support German war-related animal supplies. The officials in the Ministry of the Interior occupied with veterinary matters, Friedrich Weber and Friedrich Müssemeier, were both in favour of maintaining the OIE as an institution to restart a regular exchange of disease communication. Müssemeier, who since 1935 was the director of the veterinary department at the Reichsgesundheitsamt (Reich Health Office) and since 1941 the head of the new department for epizootics control in the Ministry of the Interior, was, in contrast to Weber, not known as a fervent National Socialist. He had longtime ties to the OIE, having been already present at the OIE founding conference in Paris in 1921, and since 1936 being the German representative at the OIE committee.[102] In a meeting at the Ministry of the Interior in December 1941, Müssemeier argued that the OIE's work was important for Europe and provided the experience and necessary infrastructure for the required tasks.[103] The Supreme Command of the Wehrmacht did not object to the publication of German disease statistics as long as they showed only favourable trends.[104] The German Foreign Office, however, argued in a note to the Ministry of the Interior that it was its definite aim to advance Germany's role as host country of international organisations, and therefore reduce the number of organisations in France and Belgium that had been founded during the 1920s and 1930s. The Foreign Office therefore was strictly against a revival of the OIE in Paris, and demanded the creation of a new international organisation in Berlin "under German leadership".[105] In contrast, both Weber and Müssemeier argued that Germany's cooperation with the OIE had worked excellently since its foundation. A note on the topic provided in early 1942 by the Ministry of the Interior outlined that the present existence of the OIE was in any case considered only of a provisional nature: after the (victorious) war and the following reorganisation of Europe according to German plans, the Ministry intended to transfer the office permanently to Germany.[106] In the current situation, the note argued, securing the war supplies was essential, and fast and effective approaches towards animal diseases should take priority over general political interests; the Ministry officials therefore asked the Foreign Office to postpone the question of the OIE transfer.[107] In April 1942, the Foreign Office indeed consented to temporarily postponing the question of the OIE's relocation to Berlin.[108] Furthermore, the financial contributions of Germany for the years 1939, 1940 and 1941 were transferred to the OIE.[109]

The infiltration, control and domination of the OIE's information policy and infrastructure by the German authorities became visible in the content of the Bulletin during the war years. Despite all problems, the publication continued to appear. Increasingly, it mirrored a German dominated version of veterinary internationalism, filled predominantly with information from German researchers and news from the German-controlled areas or client states.[110] The majority of reports on veterinary topics were based on scientific publications by German-speaking authors. Also, reports and news from Italy and Japan, from occupied territories, from Vichy France, and from neutral countries such as Switzerland were included. Still, and not surprisingly, the Bulletin appeared more like a translated overview of German veterinary activities and research in times of war, with carefully selected contents, than a publication of an international dimension. Apart from providing contributions to the Bulletin, the German authorities aimed to foster the links between the OIE and German researchers and continued to use the OIE as a platform of scientific exchange. In September 1942, for example, Otto Waldmann held a lecture at the OIE about his vaccination research against foot-and-mouth disease at Riems; he also published several articles in the OIE Bulletin.[111] Waldmann had long-term professional connections to the OIE, having acted as one of the German representatives at several of the OIE committee meetings in the late 1920s and during the 1930s. He also knew Flückiger, since he had provided technical advice in 1941 for the development of the new Swiss Federal Institute for the production of a foot-and-mouth disease vaccine in Basel (Eidgenössisches Vakzine-Institut, opened in 1942).[112]

At the same time, a separate National Socialist version of veterinary internationalism started to take a direction away from that of the OIE. In September 1942, Weber wrote to the German Foreign Office that, in his view, it was essential to advance new activities against epizootics in conformity with German interests. Germany imported high numbers of livestock especially from South-Eastern Europe, and several outbreaks of epizootics had already endangered this line of supply. Weber suggested an intergovernmental working association of the heads of veterinary administrations of several states, in particular the Slovak Republic, Hungary, Croatia, Bulgaria, Romania, the Serbian Residual State and possibly Turkey.[113] The governments of these client states, protectorates, occupied territories or neutrals signaled their interest.[114] Apart from the neutral states, most governments did certainly not have much choice in the matter. Weber initiated the organisation of a conference for the foundation of the working association.[115] He suggested the Slovak Republic (founded in 1939 as a client state of Germany) as host country because of its central geographical position. Weber's draft invitation letter to the governments indicated the close economic interrelations between importing and exporting countries as the main reason for the project. It was the aim of the initiative, he argued, to "seek, through trustful cooperation, a way that makes it possible to contain or even eradicate the most dangerous epizootics for livestock".[116] Turkey was eventually not invited because it had not officially recognised the Slovak Republic.[117] The organisers decided not to include Italy either, as Germany's arrangements with the southern European states should be the focus first.[118]

In June 1943, the Tagung zur Gründung einer zwischenstaatlichen Arbeitsgemeinschaft auf veterinärem Gebiet (Conference for the Foundation of an Intergovernmental Working Association in the Field of Veterinary Medicine) took place in Bratislava (Pressburg).[119] The participants were Weber and Müssemeier, three additional German veterinary officials,[120] and the heads of the veterinary authorities of the Slovak Republic, the Kingdom of Hungary, Croatia, and of the German Military Commander of the Serbian Residual State; the Romanian representative had excused himself.[121] The programme had a main focus on problems of the import and export of animals and animal products, shaped by the war-related significance of the topic for the German initiators.[122] Weber, who was the driving force behind the project, emphasised once again in his inaugural speech the increasingly close economic relations between central and south-eastern Europe. The future-oriented project of closer cooperation on animal diseases, Weber explained, intended to initiate steps toward greater uniformity and cooperation in the work against epizootics.[123] Weber had originally envisaged an agreement or convention, but the conference eventually ended with a final protocol with three technical attachments which established the foundation of the intergovernmental working association.[124] The German veterinary administration consented to act as the managing agency. The content of the protocol mirrored in many aspects the debates of veterinary internationalism of the last decades, also the efforts of the League of Nations and the OIE in veterinary matters since 1927, which had been cut by the war and the German occupation. It included the principal themes and aims of the working association—the establishment of a regular information service on epizootics, for certain diseases also at short notice via telegraph, the aim of creating uniform standards of communication, and an agreement on the necessity of using unified veterinary methods of organisation and control, both of animals as well as of animal products. It also pointed out the goal of establishing common regulations for animal transport by rail and even the possibility of a mutual project concerning vaccines and other medical products.[125] The participants agreed to first of all assess the situation of the veterinary organisation of their respective states down to the local level and communicate this to the other participants. A list of relevant epizootics and detailed arrangements for communication matters were included as attachments, as well as principles for the organisation of state veterinary services and meat inspection.[126] The existing or previous international arrangements and projects of the League and the OIE, which were clearly the role models for the arrangement, were not mentioned. Weber highlighted in his closing remarks of the conference the singularity and extraordinary success of the event.[127] Of course, an open discussion was not possible. The dictatorial role of the representatives of the German National Socialist government present and the pressure of the political situation made the participants agree in short time on the arrangements and their details, most of which were anyways not new. According to Weber's announcement, intergovernmental cooperation was necessarily to be based on the willingness of the cooperating partners and on "trustful cooperation".[128] The steps of the German National Socialist government toward veterinary internationalism, aiming at a new German-dominated initiative on veterinary cooperation for Central and South-Eastern Europe, could indeed not completely forgo the idea of some form of 'organising trust', but thoroughly intertwined it at all levels with dictatorial control and political-military pressure—Germany was dependent on the supplies of livestock and animal products from its occupied or client countries, as well as on their adequate measures to contain epizootics. All invited governments accepted the protocol and received invitations to the next conference, which was to take place in September 1944 in Gmunden in Upper Austria. In August 1944, however, the meeting was cancelled because of the war situation.[129]

Planning post-war international approaches to animal health

In Paris, the OIE managed to survive as an institution the German occupation. The staff had to be dismissed; only Leclainche and the general secretary Le Louet remained.[130] In August 1943, Leclainche described the situation of the OIE to the Vichy Ministry of Foreign Affairs as being more or less calm. A German veterinary officer of the Wehrmacht was present in the building.[131] Four governments (of Germany, Belgium, Romania, and Switzerland) paid their annual fees for 1943.[132] The capacity of the OIE to work was restricted by the ongoing war and its blockade of communication. Flückiger managed to remain in contact with the OIE in Paris via the Swiss Chamber of Commerce in Paris.[133] The OIE Bulletin appeared in a shorter version with statistics and scientific contributions mainly from German, French, Spanish and Portuguese veterinarians. For Leclainche and Flückiger, as they explained after the war, the key to saving the OIE as an institution was the continuation of its work and especially its publications even in isolation.[134] After the liberation of Paris in late August 1944, the staff of the OIE was confronted with new problems. In the immediate weeks afterwards, as Leclainche wrote to Flückiger, reproaches were made against him for collaboration, because he had reopened the OIE under German occupation and furthermore advocated the creation of a new French foot-and-mouth disease institution for vaccine production, which would, if realised, have necessitated a German-French cooperation. Leclainche could, as he outlined, invalidate the reproaches, explaining that the OIE was an international organisation in which he as executive director followed the directives of the OIE committee president and the committee decisions. Furthermore, he stated, he had intended to make France independent of the Riems vaccine.[135]

However, the institutional problems of the OIE continued on a different level. It soon became clear that the fight against epizootics would become a part of post-war economic reconstruction plans, integrated in a globally oriented system of international organisations and initiatives in context of the new United Nations. For the OIE, relations with the rest of the world were difficult to rebuild, even more so since the financial means of the office continued to decrease in the months after the liberation of Paris. The OIE directorate tried to make contact with Soviet and U.S.-American authorities to establish regular lines of communication. Leclainche thought that this was the moment to try to finally include the United States in future OIE work.[136] Most importantly for the institutional situation and future of the OIE as well as for the development of approaches against animal diseases on a global scale, a number of new international organisations were founded whose aims and working plans overlapped in parts with those of the OIE. They were the United Nations Relief and Rehabilitation Administration (UNRRA), a transitional institution for organising relief for the victims of war in the areas formerly occupied by the Axis powers, and, after the Hot Springs Conference in May 1943, the United Nations Interim Commission on Food and Agriculture. In October 1945, the latter led to the foundation of the Food and Agriculture Organization of the United Nations (FAO) with the aim of securing and promoting worldwide food supply, improving agricultural production and the living conditions of the rural population.[137] During this period, many pre-war international organisations were closed down or reorganised. The League of Nations was dissolved and its properties taken over by the UN in April 1946. In August 1946, the IIA was taken over by the new FAO. In July, the OIHP in Paris was dissolved.[138] For several years, it was unclear whether the OIE would continue to exist as an international institution or whether it would be incorporated into the new United Nations system.

UNRRA was founded in November 1943, following an initiative of the U.S. government, with the aim to provide relief to the victims of the war.[139] In the field of international public health, UNRRA mainly worked to prevent war-related epidemics and endemic diseases, famines and malnutrition, and provide medical and humanitarian care for so-called displaced persons.[140] UNRRA also took charge of revising international sanitary conventions and of epidemiological bulletins, giving the reason that, since the German occupation of Paris, the OIHP was no longer capable of performing its duties.[141] UNRRA aimed to quickly raise the livestock production for food and milk supply in order to re-establish destroyed economies and trade relations. This also involved activities against animal diseases, since livestock was an important part of the supplies organised by UNRRA for the war victims.[142] Given these new institutions on the stage of international work, for the OIE it was necessary to establish and re-establish international contacts soon, and to demonstrate its rationale and legitimacy in the rapidly transforming international context. During the first half year of 1945, Flückiger as OIE president, Leclainche as director, by now in his mid-eighties, and Le Louet as general secretary worked to establish a cooperation with UNRRA. The OIE directorate learned that UNRRA, similar to its activities in international health, had since autumn 1944 already been planning a new international convention on animal diseases that would also include the foundation of a new central veterinary bureau.[143] Leclainche contacted UNRRA officially via the French Ministry of Foreign Affairs and sent a memorandum about the work and current situation of the OIE.[144] Leclainche was successful—UNRRA abandoned the idea of an international convention.[145] Several UNRRA officials visited the OIE in May 1945, obviously wanting to convince themselves that the OIE still existed and was working.[146] In June, the UNRRA Standing Technical Sub-Committee on Agriculture decided to cooperate with the OIE and appointed a liaison officer to maintain contact.[147] A major problem for all sides was the ongoing difficulties of communication, which made the efforts toward international documentation and coordination extremely difficult, both within and beyond Europe.[148] Nevertheless, it turned out to be more and more essential for the OIE to convene a regular meeting of as many members of its international committee as possible to confirm the institutional existence and working capacity of the OIE and its governing body.

In July 1945, the OIE directorate first convened an extraordinary meeting of committee delegates from several neighbouring countries (Great Britain, Belgium, Switzerland, and France).[149] The meeting documented the efforts of the OIE to keep pace with the complex and overlapping global developments in the aftermath of the war. The agenda encompassed two main points: organising the resumption of regular work of the OIE, and the discussion of (past and current) foot-and-mouth disease outbreaks and their future prevention. Leclainche suggested taking up the project on foot-and-mouth disease vaccination from 1939 and starting a new international campaign in Europe:

> A country does not have the right to act as it sees fit because its neighbours are the victims of its negligence.... Therefore, from now on, a general policy of foot-and-mouth disease and its prophylaxis needs to be established for the entire European continent.[150]

Leclainche's idea was certainly to demonstrate the OIE's working capacity beyond the tasks of information transfer and animal disease documentation. But in the post-war situation of international work, the strong European orientation of the OIE's work, which again came to the fore in this project, became now a serious problem for its existence and legitimacy. UNRRA had, from the beginning, envisaged a more global and direct approach of providing agricultural relief, with a special focus on interventions in regions of both Europe and Asia where the loss of livestock due to the war was immense. In China especially, epizootics such as rinderpest were an enormous problem in a situation where pre-war sanitary and veterinary structures had mostly been destroyed.[151] Despite the work of the liaison officer, the communication between UNRRA and OIE was not always functioning well. As he reported, Leclainche learned from the newspapers in December 1945 that the South-West Pacific Area section of UNRRA had held an international veterinary conference in Sydney.[152] The conference had a focus on the Pacific region and assembled representatives from the U.S.-American and British military as well as from the United States, Australia, New Zealand, India, the Dutch East Indies, China and the Philippines. The meeting did not only deal with the organisation of restocking the herds in the Asian and Pacific region, it also created a provisional international body for the Far East, the South West Pacific Area Livestock and Veterinary Secretariat, located in the Sydney office of UNRRA. The organisation was similar to the former Singapore Office of the League of Nations Health Organisation, but with a focus on veterinary matters.[153] The OIE directorate saw the office increasingly coming under pressure. The institutional survival was no longer a question of merely continuing the office's work in the pre-war structural, thematic and conceptual framework. The challenge now was to thoroughly adapt and redefine the role of the OIE to the post-war situation and to demonstrate its relevance, also with regard to the global contexts of animal diseases.

When FAO began its activities, the landscape of international organisations in the field of animal disease control transformed profoundly and permanently. After meeting informally in New York in June 1946, a group of veterinary officials under the umbrella of the FAO held a meeting on the topic of animal health in London in August 1946.[154] The emphasis of the FAO initiative was to address the dramatic worldwide food shortages, and from this perspective, a more concerted and active global approach towards epizootics was highlighted as a central issue, especially with regard to rinderpest. At the London meeting, the participants discussed the relations of FAO with the various organisations and the coordination of their work on animal diseases. Apart from FAO, listed were now the Imperial Agricultural Bureaux, the international veterinary congresses, the OIE, and the UNRRA and its Southwest Pacific Area Secretariat (until 1947). A new regional Central American-Mexican Office of Animal and Vegetable Sanitation was founded in early 1947 in San Salvador. As representatives of the OIE, both Leclainche and Flückiger were invited and took part in the meeting in London.[155] The question of a transfer of the OIE to the FAO, as had already happened in the case of the IIA, was indeed open. The meeting in London ended with the FAO representatives supporting the continuation of the OIE, linked to the demand for an extension of OIE activities and its cooperation with other international organisations. A meeting in context of the FAO conference in Copenhagen in September 1946 confirmed this position.[156]

In October 1946, the members of the OIE international committee finally convened for a new session, the first one since 1939. It included representatives of 26 member states, mostly Europeans because of the travel difficulties. Turkey, the Soviet Union, and several colonial states were represented (by their colonial power); an Austrian representative took part again.[157] In his report on the sanitary situation of 1946 included in the OIE Bulletin, Leclainche argued that the observable more pluralised and complex patterns of animal disease appearances required a profound change in the office's orientation towards a more comprehensive approach on connected national and international levels. New questions of animal health and hygiene, going beyond previous approaches against epizootics, needed to be formulated and negotiated. The OIE had, Leclainche outlined, already envisaged such a broad conceptual approach towards animal health during the 1930s:

> The 'Office of Epizootics' must be an 'Office of Animal Health', with the task of elaborating and codifying the rules of preventive and applied hygiene, whether with the assistance and under the control of administrative authorities, or directly by those concerned themselves, collectively or individually.[158]

The OIE committee meeting in October 1946 was therefore a demonstration of the active resumption of the office's work and its function as a platform for expert discussions, as well as of its legitimacy to continue as an independent institution under the new international conditions. Flückiger, who was confirmed as OIE committee president, emphasised as well the importance of expanding the fields of activity and the new tasks, which, as he outlined, required securing funding and a reorganisation of the OIE. The office's financial resources were exhausted, but the institution still owned its office building in Paris.[159] The meeting passed resolutions on foot-and-mouth disease and dourine, and strengthened the call for an international approach to foot-and-mouth disease vaccination.[160] Reporting on the conference of the FAO in London, Leclainche outlined that the future focus of the OIE required an extension of its programme. This concerned two factors: a new orientation on generally all economically and public health-related questions of animal health, and structurally, close cooperation with other organisations occupied with questions of agriculture and public health.[161] This direction pointed not only to a stronger focus on enzootic and non-contagious animal diseases, but also to the many new or re-emerging challenges posed by zoonotic diseases, and it also included a more comprehensively global perspective.

The future of the OIE was, however, dependent on negotiations with the FAO. In March/April 1947, the new Subcommittee on Animal Health of the FAO Standing Advisory Committee on Agriculture met in Washington to begin the coordination of international veterinary work.[162] One of the points of the agenda was to examine the relationship with other international veterinary organisations, among them the OIE. Flückiger took part as OIE representative. In a report produced for the meeting, probably authored by him, he argued that the OIE was flexible for future adaptations of its agenda:

> A discreet but powerful influence may be exercised upon Governments often slow in their regulations and interventions, as well as solutions common to several countries may be reached by examining together the same questions. In so doing the Office efficiently paves the way to this international collaboration which is so much longed for.[163]

The office had, the report further outlined, in any case always worked on the extension of its tasks; therefore, its activities went far beyond its narrow title and encompassed multiple areas of animal health—not only epizootics, but all questions of economic interest. The report highlighted that a new world health organisation needed an equivalent on animal health to provide the necessary inputs and exchange.[164] The participants of the meeting discussed different possible scenarios of the relationship between FAO and OIE. Ideas to fuse or affiliate the OIE with FAO were dropped due to legal issues, such as different national memberships.[165] The participants decided eventually to keep the independence of the OIE as a separate institution, and they underlined the significance of establishing an agreement on close and constant forms of cooperation with FAO to avoid overlapping projects and duplicating functions.[166] The subsequent OIE committee meeting in May 1947 discussed at length the future relationship and the draft of an arrangement with the FAO.[167] Sir Daniel Cabot, Chief Veterinary Officer for the United Kingdom, since 1946 vice-president of the OIE committee, was present as FAO representative. In 1949, Cabot would become Flückiger's successor as OIE committee president. As Leclainche pointed out, the appointment of Cabot already showed the goodwill of FAO towards the OIE.[168] Leclainche argued strongly for building a constructive and loyal relationship with FAO. This was, he stressed, necessary not only because of the OIE's lack of interventive powers, but also to advance the global dimension and a more comprehensive approach of the work against animal diseases. The office needed, in his view, the cooperation with FAO also to expand its network to provide updated disease information.[169] The meeting discussed the text of an agreement note between the FAO and the OIE, based on the points outlined at the FAO subcommittee meeting in Washington. The agreement text defined the competencies of the two organisations as well as their mutual support, communication, exchange of information and cooperation. It underlined the established position of the OIE as an international organisation to address the fight against animal diseases. Further, it included a paragraph on the extension of the OIE's scope of activities: in future, these were to focus more strongly on the aspect of animal disease prevention. To this end, the text integrated plans of the OIE to establish regional offices in non-European countries.[170] The discussion on this point included the idea of an office in Saigon, as well as of taking over the UNRRA veterinary secretariat in Sydney after UNRRA's dissolution.[171]

Indeed, in September 1947, the negotiations between the OIE and the FAO resulted in a note of understanding between the institutions to coordinate their future relationship. The note enhanced that the "magnitude and diversity of the problem are such that full cooperation among national and international agencies is essential to avoid duplication and dissipation of effort".[172] The future tasks of the OIE were defined by reference to its founding agreement of 1924 with inclusion of some updates. The first point concerned the promotion of research and experiments concerning infectious and—newly—parasitic diseases of domestic animals. The second referred to the traditional core task of the office, the collection and distribution of information concerning epizootics and anti-disease measures to the member governments. The last point underlined the task of the OIE to prepare international agreements for member governments.[173] The note outlined that the FAO should foster cooperative international activities with regard to the training of veterinarians and the establishment of veterinary schools especially in poor countries, collect information on the situation of veterinary administrations with regard to equipment and supplies, work on the information on the cross-border movements of animals (especially with regard to the question of quarantine in case of breeding animals), the provision of veterinary facilities with research literature, and advance cooperative research efforts in the field of animal diseases. Important was the last point of its tasks, which mentioned a more active role towards global animal health: "take steps, as the need arises, to encourage and facilitate developments designed to protect animal health", such as in the case of rinderpest and the laboratory cooperation in vaccination efforts.[174] The note further underlined the FAO's recognition of the OIE and the areas of cooperation, namely by sending representatives to meetings and a regular exchange of information on activities, as well as encouraging FAO member states to join the OIE.[175]

With the distribution of tasks and its institutional basis solidified, at least for the moment, the OIE committee began planning new activities. In the case of rinderpest, it was the FAO that integrated large-scale efforts of research and disease control on site through vaccination as a central part of its animal health programme. In the resolutions of its meeting of May 1947, the OIE committee underlined the significance of the recent approaches against rinderpest on a global scale and announced its support in connection with the efforts of FAO and British colonial authorities.[176] FAO was, in the following decades, the main institutional player in the new international fight against rinderpest.[177] In the case of foot-and-mouth disease, the OIE was able to take up earlier plans of activities, at least with regard to Europe. In October 1946, the OIE committee meeting had agreed on a resolution to advance the fight against foot-and-mouth disease, calling again for international cooperation to organise the production and distribution of a vaccine.[178] A small international meeting of specialists in Berne in November 1946[179] resulted in an arrangement on the production and distribution of foot-and-mouth disease vaccines. It expressed the aim of an internationally organised supply of vaccines coordinated by the OIE.[180] As a next step, an international conference convened in September/October 1947, again in Berne, calling for a new "entente internationale" to tackle foot-and-mouth disease.[181] However, the position of the OIE remained very difficult in the following years, and its capacities beyond the organisation of information and documentation to launch activities to address animal diseases in their transboundary dimension were very small. Despite the note of understanding of 1947, the cooperation between FAO and OIE was not without difficulties in the years that followed—FAO developed its globally oriented veterinary work largely without the OIE. As Flückiger outlined in a synthesis paper on international veterinary efforts published in 1954, the cooperation was limited to the mutual representation at the respective conferences.[182] Flückiger added that several other new international organisations began activities addressing animal disease problems, among them the WHO, founded in 1948, which soon began to organise a focus on zoonoses.[183] Other newly founded organisations that addressed animal disease problems included, in Europe, the newly founded Organisation for Economic Cooperation in Europe (OEEC), and in the Americas, the Pan American Foot-and-Mouth Disease Center (PANAFTOSA).[184] The independence of the OIE, directed since 1948 by Leclainche's successor Gaston Ramon, remained at risk. Flückiger outlined that the plans to integrate the OIE into the FAO were considered again in 1950, but could be avoided. The problem of overlapping international structures and rivalling activities remained.[185] In 1952/53, FAO and OIE finally agreed on a formal agreement between the two organisations to coordinate future cooperation. FAO hereby recognised the OIE as an advisory body on animal disease issues.[186] The OIE began also to focus increasingly on zoonotic diseases, in cooperation with other international organisations. Added to the agreement with the FAO was an official cooperation with the WHO in 1960. Several decades later, in 2003, the OIE changed its name to the "World Organisation for Animal Health".[187] Next to rinderpest, which was declared eradicated in 2011, the idea of addressing foot-and-mouth disease on a comprehensive and global level has continuously been a central issue of both FAO and OIE. With a conference in Asunción in 2009, the OIE and FAO launched the Global Foot-and-Mouth Disease Control Strategy as a comprehensive joint transboundary programme against foot-and-mouth disease, and elaborated on its work in more detail at a second conference in 2012 in Bangkok. The initiative aims to control foot-and-mouth disease globally and at the same time support animal health systems and the related infrastructure on a global level. The final and long-term aim is to eradicate foot-and-mouth disease globally—should it succeed, this would be the second animal disease to be eliminated by efforts of international cooperation to advance animal health.[188]

Notes

1 Sécrétariat de la Commission Permanente des Congrès Internationaux de Médecine Vétérinaire, signed de Blieck, to the Deutsche Kongress Zentrale, March 17, 1939. Hoover Institution Archives: Germany, DKZ, Box 307 (Tierärzte-Kongresse). See also: Hans Stümke, ed., *Bibliographie der internationalen medizinischen Kongresse und Verbände*, Bibliographie der internationalen Kongresse und Verbände 1 (Berlin: O. Harrassowitz, 1939), 218–219.

2 About the history of the OIE: Peter Ludwig, *Die Arbeiten des Internationalen Tierseuchenamtes in Paris seit seiner Gründung im Jahre 1921 bis zum Ausbruch des europäischen Krieges im September 1939* (Bern: J. Werder, 1941); Marie Teissier, 'A Brief History of the OIE', *OIE Bulletin*, no. 1 (2007): 6–18.

3 République Française, Ministère de l'Agriculture, ed., *Conférence Internationale pour l'Etude des Epizooties. Paris (25–28 Mai 1921). Compte Rendu Sténographique* (Paris: Imprimerie Nationale, 1921), Annexe, Statuts, 3 (Art. 4).

4 Procès-Verbaux de la Première Réunion du Comité de l'Office International des Epizooties, in: *OIE Bulletin* 1 (Juillet 1927–Juin 1928), 14–25.

5 Procès-Verbaux de la Première Réunion du Comité de l'Office International des Epizooties, in: *OIE Bulletin* 1 (Juillet 1927–Juin 1928), 15, 24.

6 Liste des Pays Participant à l'Office International des Epizooties, in: *OIE Bulletin* 1 (Juillet 1927–Juin 1928), IV.

7 About the OIE membership and the European focus see also: Mark Harrison, *Contagion: How Commerce Has Spread Disease* (New Haven: Yale University Press, 2012), 227.

8 In response to an inquiry of the government of the URSS in May 1935, the OIE secretariat established an updated detailed list of the status of current and former members. OIE, Leclainche à Monsieur le Ministre des Affaires Etrangères (France), May 29, 1935, attached: "Liste des Puissances". ADLC: Unions, 429QO/335 (Office International des Epizooties, 1934–39).

9 OIE, Leclainche à Monsieur le Ministre des Affaires Etrangères (France), July 5, 1931, 1. ADLC: Unions, 429QO/334 (Office International des Epizooties, 1929).

10 OIE, Leclainche à Monsieur le Ministre des Affaires Etrangères (France), February 2, 1933, 3. ADLC: Unions, 429QO/334 (Office International des Epizooties: Contributions 1929 et seq./Grande Bretagne).

11 OIE, Leclainche à Monsieur le Ministre des Affaires Etrangères (France), February 2, 1933, 3–4. ADLC: Unions, 429QO/334 (Office International des Epizooties: Contributions 1929 et seq./Grande Bretagne).

12 British Embassy Paris to the Ministère des Affaires Etrangères, July 8, 1932. ADLC: Unions, 429QO/334 (Office International des Epizooties: Contributions 1929 et seq./Possessions Britanniques); OIE: Leclainche à Monsieur le Ministre des Affaires Etrangères (France), February 2, 1933, 3. ADLC: Unions, 429QO/334 (Office des Epizooties: Contributions 1929 et seq./Grande Bretagne).

13 'Imperial Bureaux', *The British Medical Journal*, No. 3546 (Dec. 22, 1928), 1148–9.

14 Liste des Pays Participant à l'Office International des Epizooties, in: *OIE Bulletin* 17 (Juillet 1938–Avril 1939): VI.

15 Gottlieb Flückiger, 'Bisherige Bemühungen zur internationalen Bekämpfung tierischer Seuchen', *Schweizer Archiv für Tierheilkunde* 96, no. 1 (1954): 18. Gottlieb Flückiger (1892–1987), a specialist on epizootics and veterinary public health, was since 1932 the successor of Moritz Bürgi as head of the Swiss federal veterinary administration and served in that function until 1957. In 1939 he was elected OIE committee president as successor of Berger. About Flückiger see: Therese Steffen Gerber, 'Flückiger, Gottlieb', Historisches Lexikon der Schweiz, accessed March 19, 2021, http://www.hls-dhs-dss.ch/textes/d/D17457.php.

16 OIE: Leclainche à Monsieur le Ministre des Affaires Etrangères (France), October 19, 1938. ADLC: Unions, 429QO/357 (Evacuation des Organismes Internationaux en Cas de Guerre).

17 Berger, in: Huitième Session du Comité de l'Office Mai-Juin 1934, Procès-Verbaux des Séances, in: *OIE Bulletin* 8 (Mai-Juin 1934), 455.

18 Berger, in: Compte-Rendu de la Dixième Session du Comité de l'Office, Juin 1936, in: *OIE Bulletin* 12 (1936), 499.

19 About the OIE building: Rapport Administratif du Directeur de l'Office pour l'Année 1938–1939, Treixième Session du Comité de l'Office, Mai-Juin 1939, in: *OIE Bulletin* 13 (1939), 351–2.

20 République Française, Ministère de l'Agriculture, *Conférence Internationale pour l'Etude des Epizooties 1921*, Annexe, Statuts, 3 (Art. 2).

21 République Française, Ministère de l'Agriculture, *Conférence Internationale pour l'Etude des Epizooties 1921*, Annexe, Statuts, 3 (Art. 4). Translation: World Organisation for Animal Health (OIE), 'Appendix to the International Agreement: Organic Statutes of the Office International Des Epizooties', accessed April 1, 2021, https://www.oie.int/about-us/key-texts/basic-texts/organic-statutes/.

22 Bürgi, in: Cinquième Session du Comité de l'Office, Procès-Verbaux des Séances, *OIE Bulletin* 5 (May 1931–Avril 1932), 194 (translation: CK). Moritz Bürgi (1878–1932) was since 1915 head of the Swiss Federal Veterinary Office (Eidgenössisches Veterinäramt) and had a significant influence on Swiss veterinary approaches against epizootics; see: Therese Steffen Gerber, 'Bürgi, Moritz', Historisches Lexikon der Schweiz, accessed March 19, 2021, http://www.hls-dhs-dss.ch/textes/d/D17456.php.

23 Emmanuel Leclainche, 'La "Standardisation" des Bulletins Sanitaires', *OIE Bulletin* 1 (Juillet 1927–Juin 1928), 577–83.

24 Leclainche, 'La "Standardisation" des Bulletins Sanitaires', 581–2.

25 Leclainche, in: Cinqième Session du Comité de l'Office International des Epizooties, Procès-Verbaux des Séances, in: *OIE Bulletin* 5 (Mai 1931–Avril 1932), 200.

26 Leclainche, in: Sixième Session du Comité de l'Office International des Epizooties, Procès-Verbaux des Séances, in: *OIE Bulletin* 6 (Mai 1932–Avril 1933), 210.

27 About the problems of data with the example of foot-and-mouth disease: Septième Session du Comité de l'Office International des Epizooties, Procès-Verbaux des Séances, in: *OIE Bulletin* 7 (Mai 1933–Avril 1934), 277.

28 Leclainche, in: Cinqième Session du Comité de l'Office International des Epizooties, Procès-Verbaux des Séances, in: *OIE Bulletin* 5 (Mai 1931–Avril 1932), 196.

29 Leclainche, in: Sixième Session du Comité de l'Office, Procès-Verbaux des Séances, in: *OIE Bulletin* 6 (Mai 1932–Avril 1933), 210–11.

30 Leclainche, in: Septième Session du Comité de l'Office International des Epizooties, Procès-Verbaux des Séances, in: *OIE Bulletin* 7 (Mai 1933–Avril 1934), 275.

31 Emmanuel Leclainche, 'Un Concept d'une Police Sanitaire Moderne', *OIE Bulletin* 6 (Mai 1932–Avril 1933), 173–4.

32 Leclainche, 'Un Concept d'une Police Sanitaire Moderne', 175 (translation: CK).

33 Leclainche, 'Un Concept d'une Police Sanitaire Moderne', 176.

34 For an overview on the general sessions of the OIE committee and their topics see also the summaries provided by Ludwig, *Die Arbeiten des Internationalen Tierseuchenamtes*.

35 Madeleine Herren, 'Fascist Internationalism', in *Internationalisms. A Twentieth-Century History*, ed. Glenda Sluga and Patricia Clavin (Cambridge: Cambridge University Press, 2017), 203–6; Elisabetta Tollardo, *Fascist Italy and the League of Nations, 1922–1935* (London: Palgrave Macmillan, 2016), 34–7.

36 Leclainche, in: Sixième Session du Comité de l'Office International des Epizooties, Procès-Verbaux des Séances, in: *OIE Bulletin* 6 (Mai 1932–Avril 1933), 212.

37 Leclainche, in: Septième Session du Comité de l'Office International des Epizooties, Procès-Verbaux des Séances, in: *OIE Bulletin* 7 (Mai 1933–Avril 1934), 273–4.

38 Leclainche, in: Septième Session du Comité de l'Office International des Epizooties, Procès-Verbaux des Séances, in: *OIE Bulletin* 7 (Mai 1933–Avril 1934), 273–4 (translation: CK).

39 See Chapter 5.

40 Art. 5, in: United Nations Treaty Collection, 'International Convention for the Campaign against Contagious Diseases of Animals. Geneva, February 20th, 1935', United Nations Treaty Collection, accessed February 19, 2021, https://treaties.un.org/Pages/LONViewDetails.aspx?src=LON&id=568&chapter=30&clang=_en (referring to: League of Nations, Treaty Series, vol. 186, 173–90, here 182–3).

41 Deuxième Réunion du Comité de l'Office, Procès-Verbaux des Séances, in: *OIE Bulletin* 1 (Juillet 1927–Juin 1928), 607–8.

42 About the League's efforts of standardisation of biological products see: Pauline M. H. Mazumdar, '"In the Silence of the Laboratory": The League of Nations Standardizes Syphilis Tests', *Social History of Medicine* 16, no. 3 (2003): 437–59; Iris Borowy, *Coming to Terms with World Health. The League of Nations Health Organisation, 1921–1946* (Frankfurt am Main and New York: Peter Lang, 2009), 114–16, 154–60.

43 Borowy, *Coming to Terms with World Health*, 114–16, 154–60.

44 Leclainche, in: Sixième Session du Comité de l'Office, Procès-Verbaux des Séances, in: *OIE Bulletin* 6 (Mai 1932–Avril 1933), 212.

45 Note sur la Standardisation des Produits Biologiques Utilisés dans la Medicine Vétérinaire, in: Sixième Session du Comité de l'Office, Procès-Verbaux des Séances, in: *OIE Bulletin* 6 (Mai 1932–Avril 1933), 213–14.

46 Note sur la Standardisation des Produits Biologiques Utilisés dans la Medicine Vétérinaire, in: Sixième Session du Comité de l'Office, Procès-Verbaux des Séances, in: *OIE Bulletin* 6 (Mai 1932–Avril 1933), 213–14.

47 Sixième Session du Comité de l'Office, Procès-Verbaux des Séances, *OIE Bulletin* 6 (Mai 1932–Avril 1933), 215.

48 Leclainche, in: Huitième Session du Comité de l'Office Mai-Juin 1934, Procès-Verbaux des Séances, in: *OIE Bulletin* 8 (Mai-Juin 1934), 458.

49 Leclainche, in: Huitième Session du Comité de l'Office, Procès-Verbaux des Séances, in: *OIE Bulletin* 8 (Mai-Juin 1934), 458; Résolutions adoptées dans la 8ème Session, 533–4.

50 Leclainche, in: Huitième Session du Comité de l'Office, Procès-Verbaux des Séances, in: *OIE Bulletin* 8 (Mai-Juin 1934), 458–9.

51 Rapport Administratif du Directeur de l'Office, in: Neuvième Session du Comité de l'Office International des Epizooties, in: *OIE Bulletin* 10 (Mai-Juin 1935), 353.

52 Rapport Administratif du Directeur de l'Office, in: Neuvième Session du Comité de l'Office International des Epizooties, in: *OIE Bulletin* 10 (Mai-Juin 1935), 353.

53 Rapport Administratif du Directeur de l'Office and comment, in: Neuvième Session du Comité de l'Office International des Epizooties, in: *OIE Bulletin* 10 (Mai-Juin 1935), 353, 367–8, 372.

54 Rapport Administratif du Directeur de l'Office, in: Onzième Session du Comité de l'Office, *OIE Bulletin* 14 (Mai-Juin 1937), 364.

55 Onzième Session du Comité de l'Office, *OIE Bulletin* 14 (Mai-Juin 1937), 479–80; see also: Ludwig, *Die Arbeiten des Internationalen Tierseuchenamtes*, 53–4.

56 Emmanuel Leclainche, 'Die derzeitige Verbreitung der Maul- und Klauenseuche und die Möglichkeiten ihrer Prophylaxe', in *XVIII. Internationaler Landwirtschaftskongress Dresden, 6. bis 12. Juni 1939*, vol. 3, Sektion VI: Tierzucht und Tierhaltung (Berlin: Reichsnährstand Verlags-Ges.m.b.H, 1939), 20.

57 Rapport Administratif du Directeur de l'Office, Douzième Session du Comité de l'Office, in: *OIE Bulletin* 16 (Mai-Juin 1938), 258–65, about the telegraphic communication service: 259.

58 Rapport Administratif du Directeur de l'Office, Douzième Session du Comité de l'Office, in: *OIE Bulletin* 16 (Mai-Juin 1938), 259–64.

59 Rapport Administratif du Directeur de l'Office, Douzième Session du Comité de l'Office, in: *OIE Bulletin* 16 (Mai-Juin 1938), 261.

60 Circular Letter to the OIE Delegations, in: Rapport Administratif du Directeur de l'Office, in: *OIE Bulletin* 16 (Mai-Juin 1938), 262–3.

61 In 1943, the institute became part of direct federal administration and was called Reichsforschungsanstalt Insel Riems (Reich Research Institute Riems Island); today it is the Friedrich-Loeffler-Institute. See: Friedrich-Loeffler-Institut (FLI), 'Friedrich-Loeffler-Institut: Historie – Standort Insel Riems', accessed January 31, 2021, https://www.fli.de/de/ueber-das-fli/historie/standort-insel-riems/.

62 On the research on foot-and-mouth disease vaccination at Riems 1933–45: Jan Ulrich Lichte, 'Die Forschung auf der Insel Riems von 1933 bis 1945 unter besonderer Berücksichtigung der NS-Zwangsarbeiter' (PhD Thesis, Ernst-Moritz-Arndt-Universität Greifswald, 2011), here 34–8.

63 The veterinarian and researcher Otto Waldmann (1885–1955) made his career as head of the Riems research institute from 1919 until the end of the Second World War and again from 1946 to 1948 with a focus on foot-and-mouth disease research; he was member of the NSDAP. About Waldmann and the history of the Friedrich-Loeffler-Institute on the island of Riems and the experiments on foot-and-mouth disease see: Friedrich-Loeffler-Institut (FLI), 'Friedrich-Loeffler-Institut: 1910–2010. FLI Riems: History and Development', accessed March 19, 2021, http://www.fli.bund.de/en/startseite/friedrich-loeffler-institut/historie/standort-insel-riems.html.

64 Experiments were conducted in 1943 at Lake Peipus in occupied Soviet territory; further tests did not take place due to the course of the war. See: Universität Greifswald, 'Die Universität Greifswald im Nationalsozialismus. Die besonderen Beziehungen zur Insel Riems', accessed May 1, 2021, https://ns-zeit.uni-greifswald.de/themengebiete/kriegseinsatz/kriegswichtige-forschungen-und-profilwandel/die-besonderen-beziehungen-zur-insel-riems/.

65 Angela von den Driesch and Peter Joris, *Geschichte der Tiermedizin. 5000 Jahre Tierheilkunde* (Stuttgart: Callwey, 2002), 186; Friedrich-Loeffler-Institut (FLI), 'Friedrich-Loeffler-Institut: Historie—Standort Insel Riems'.

66 Gottlieb Flückiger, 'Zur Frage der Abschlachtung bei Maul- und Klauenseuche', in *XVIII. Internationaler Landwirtschaftskongress Dresden, 6. bis 12. Juni 1939*, vol. 3 (Berlin: Reichsnährstand Verlags-Ges.m.b.H, 1939), 51–2.

67 *XVIII. Internationaler Landwirtschaftskongress Dresden, 6. bis 12. Juni 1939, Sektion VI: Tierzucht und Tierhaltung, Spezialberichte. Thema 1: Die neuesten Erfahrungen bei der Bekämpfung der Maul- und Klauenseuche* (Berlin: Reichsnährstand Verlags-Ges.m.b.H, 1939). About the XVIII. International Congress on Agriculture see also: Hoover Institution Archives: Germany, DKZ, Box 190 (Internationaler Landwirtschaftskongress 1939).

68 For German activities in congress internationalism during the 1930s see: Madeleine Herren, '"Outwardly… an Innocuous Conference Authority": National Socialism and the Logistics of International Information Management', *German History* 20, no. 1 (2002): 67–92.

69 Leclainche, 'Die derzeitige Verbreitung der Maul- und Klauenseuche und die Möglichkeiten ihrer Prophylaxe', 25; Flückiger, 'Zur Frage der Abschlachtung bei Maul- und Klauenseuche', 51.

70 Resolution cited in: Ludwig, *Die Arbeiten des Internationalen Tierseuchenamtes*, 67.

71 Résolutions adoptées, in: Treizième Session du Comité de l'Office, Procès Verbaux des Séances, in: *OIE Bulletin* 18 (Mai-Juin 1939), 504 (translation: CK); also: Organisation Internationale de la Vaccination contre le Fièvre Aphteuse, in: *OIE Bulletin* 19/1 (Juillet-Août 1939), 55.

72 Organisation Internationale de la Vaccination contre le Fièvre Aphteuse, in: *OIE Bulletin* 19/1 (Juillet-Août 1939), 55–7. See also: Ludwig, *Die Arbeiten des Internationalen Tierseuchenamtes*, 67–71.

73 Leclainche, 'Die derzeitige Verbreitung der Maul- und Klauenseuche und die Möglichkeiten ihrer Prophylaxe', 25.

74 Organisation Internationale de la Vaccination contre le Fièvre Aphteuse, in: *OIE Bulletin* 19/1 (Juillet-Août 1939), 57.

75 Organisation Internationale de la Vaccination contre le Fièvre Aphteuse, in: *OIE Bulletin* 19/1 (Juillet-Août 1939), 57.

76 Organisation Internationale de la Vaccination contre le Fièvre Aphteuse, in: *OIE Bulletin* 19/1 (Juillet-Août 1939), 57.

77 Organisation Internationale de la Vaccination contre le Fièvre Aphteuse, in: *OIE Bulletin* 19/1 (Juillet-Août 1939), 57–8.

78 Organisation Internationale de la Vaccination contre le Fièvre Aphteuse, in: *OIE Bulletin* 19/1 (Juillet-Août 1939), 58.

79 Ludwig, *Die Arbeiten des Internationalen Tierseuchenamtes*, 71.

80 Emmanuel Leclainche, 'L'Epizootie Aphteuse de 1937 à 1946', *OIE Bulletin* 26 (1946): 24–33.

81 OIE, Leclainche à Monsieur le Ministre des Affaires Etrangères (France, Vichy), January 28, 1941, 1. ADLC: Unions, 429QO/335 (Office International des Epizooties, 1940–1944).

82 [Ministère des Affaires Etrangères (France, Vichy)], à Monsieur le Professeur Leclainche, Directeur de l'Office International des Epizooties, December 6, 1940. ADLC: Unions, 429QO/335 (Office International des Epizooties, 1940–1944).

83 OIE, Leclainche à Monsieur le Ministre des Affaires Etrangères (France, Vichy), January 28, 1941, 1. ADLC: Unions, 429QO/335 (Office International des Epizooties, 1940–1944).

84 OIE, Leclainche à Monsieur le Ministre des Affaires Etrangères (France, Vichy), January 28, 1941, 1. ADLC: Unions, 429QO/335 (Office International des Epizooties, 1940–1944).

85 OIE, Leclainche à Monsieur le Ministre des Affaires Etrangères (France, Vichy), January 28, 1941, 1. ADLC: Unions, 429QO/335 (Office International des Epizooties, 1940–1944).
86 OIE, Leclainche à Monsieur le Ministre des Affaires Etrangères (France, Vichy), January 28, 1941, 1. ADLC: Unions, 429QO/335 (Office International des Epizooties, 1940–1944).
87 OIE, Leclainche à Monsieur le Ministre des Affaires Etrangères (France, Vichy), February 4, 1941. ADLC: Unions, 429QO/335 (Office International des Epizooties, 1940–1944).
88 Ministère de l'Agriculture (France, Vichy), Le Ministre Sécrétaire d'Etat à l'Agriculture et au Ravitaillement à Monsieur le Chef du Gouvernement [...], July 15, 1942. ADLC: Unions, 429QO/335 (Office International des Epizooties, 1940–1944).
89 OIE, Leclainche à Monsieur le Ministre des Affaires Etrangères (France, Vichy), January 28, 1941, 1 (translation: CK). ADLC: Unions, 429QO/335 (Office International des Epizooties, 1940–1944).
90 A biographical note on Friedrich Weber (1892–1954) and his career as "Reichstierärzteführer" (Reich Führer of Veterinarians, translation: CK) was established by: Deutsche Kongress-Zentrale (DKZ), Aktennotiz, Reichstierärzteführer Pg. Prof. Dr. Friedrich Weber. Hoover Institution Archives: Germany, DKZ, Box 307 (Tierärzte-Kongresse). For a biography see also: Swantje Insenhöfer, *Dr. Friedrich Weber. Reichstierärzteführer von 1934 bis 1945* (PhD thesis, Tierärztliche Hochschule Hannover, 2008), about Weber's voyages to international congresses 148–51.
91 Auswärtiges Amt (Deutsches Reich) an die Deutsche Botschaft Paris, January 18, 1941, 1–2. PAAA, Botschaft Paris 1.376 (Botschaft Paris: 246 Veterinärwesen).
92 Der Reichsminister des Innern (Deutsches Reich) an das Auswärtige Amt, signed Weber, March 11, 1941. PAAA, Botschaft Paris 1.376 (Botschaft Paris: 246 Veterinärwesen).
93 OIE, Leclainche à Monsieur le Ministre des Affaires Etrangères (France, Vichy), December 17, 1941, 1. ADLC: Unions, 429QO/335 (Office International des Epizooties, 1940–1944).
94 OIE, Leclainche à Monsieur le Ministre des Affaires Etrangères (France, Vichy), December 17, 1941, 1 (translation: CK). ADLC: Unions, 429QO/335 (Office International des Epizooties, 1940–44).
95 OIE, Leclainche à Monsieur le Ministre des Affaires Etrangères (France, Vichy), December 17, 1941, 1. ADLC: Unions, 429QO/335 (Office International des Epizooties, 1940–1944).
96 OIE, Leclainche à Monsieur le Ministre des Affaires Etrangères (France, Vichy), May 5, 1941, 1. ADLC: Unions, 429QO/335 (Office International des Epizooties, 1940–1944).
97 OIE, Leclainche à Monsieur le Ministre des Affaires Etrangères (France, Vichy), December 17, 1941. ADLC: Unions, 429QO/335 (Office International des Epizooties, 1940–1944).
98 [Ministère des Affaires Etrangères (France, Vichy)] to the Légation d'Irelande, Vichy, February 19, 1942. ADLC: Unions, 429QO/335 (Office International des Epizooties, 1940–44).
99 Ministère des Affaires Etrangères (France, Vichy), L'Amiral de la Flotte, Ministre Sécrétaire d'Etat aus Affaires Etrangères, à Monsieur le Ministre Sécrétaire d'Etat de l'Agriculture, signed Rochat, January 6, 1942. ADLC: Unions, 429QO/335 (Office International des Epizooties, 1940–44).
100 Ministère des Affaires Etrangères (France, Vichy), Circular, sent to the director of the Office International des Epizooties, November 27, 1943; OIE, Leclainche à Monsieur le Ministre Plénipotentiaire [...], Ministère des Affaires

Etrangères (France, Vichy), March 11, 1944. ADLC: Unions, 429QO/335 (Office International des Epizooties, 1940–1944).

101 L'Office International des Epizooties de 1939 à 1946, Compte Rendu de la Quatorzième Session du Comité de l'Office, in: *OIE Bulletin* 26 (1946), 7.

102 Berger, in: Compte Rendu de la Dixième Session du Comité de l'Office, in: *OIE Bulletin* 12 (1936), 495–6. The role of Friedrich Müssemeier (1876–1954) and other veterinarians in the National Socialist government and administration was illuminated during denazification trials, especially in 1948 against Weber. See, also on Müssemeier: Insenhöfer, *Friedrich Weber*, 61–3, 181.

103 [Auswärtiges Amt (Deutsches Reich)], Aufzeichnung (about the meeting on December 12, 1941), signed Navé, December 13, 1941. PAAA, R 49.159 (Akten betreffend Internationales Tierseuchenamt in Paris).

104 Der Reichsminister des Innern (Deutsches Reich) an das Auswärtige Amt, signed Weber, October 24, 1941, 2. PAAA, R 49.159 (Akten betreffend Internationales Tierseuchenamt in Paris).

105 Auswärtiges Amt (Deutsches Reich) an den Reichsminister des Innern, signed E. Albrecht, December 24, 1941, 1 (translation: CK). PAAA, R 49.159 (Akten betreffend Internationales Tierseuchenamt in Paris).

106 Der Reichsminister des Innern (Deutsches Reich) an das Auswärtige Amt, signed H. Pfundtner, January 22, 1942, 8. PAAA, R 49.159 (Akten betreffend Internationales Tierseuchenamt in Paris).

107 Der Reichsminister des Innern (Deutsches Reich) an das Auswärtige Amt, signed H. Pfundtner, January 22, 1942, 5–8. PAAA, R 49.159 (Akten betreffend Internationales Tierseuchenamt in Paris).

108 [Auswärtiges Amt (Deutsches Reich)] an den Reichsminister des Innern, April 14, 1942 (draft). PAAA, R 49.159 (Akten betreffend Internationales Tierseuchenamt in Paris).

109 [Reichsinnenministerium (Deutsches Reich)] an das Internationale Tierseuchenamt, signed Müssemeier, stamped June 26, 1942. PAAA, R 49.159 (Akten betreffend Internationales Tierseuchenamt in Paris).

110 See *OIE Bulletin* 20 (Mai 1940–Décembre 1941), 21 (Janvier-Décembre 1942), 22 (Janvier–Décembre 1943), 23 (Janvier-Mars and Avril-Juin 1944).

111 Otto Waldmann, 'La Vaccination contre la Fièvre Aphteuse', *OIE Bulletin* 21 (1942): 133–49; Otto Waldmann, 'Sur l'Aspect Futur de la Lutte contre la Fièvre Aphteuse', *OIE Bulletin* 21 (1942): 331–57. About Waldmann's lecture see also: Lichte, 'Forschung auf der Insel Riems', 54–5.

112 Flückiger (Schweizerisches Bundesveterinäramt) to Weber (Reichsministerium des Innern, Deutsches Reich), April 17, 1941. BAR, 7270(B)/1968/171, Dossier 19/2: OIE Wiederaufnahme des Büros in Paris 1941.

113 Der Reichsminister des Innern (Deutsches Reich) an das Auswärtige Amt, signed Weber, September 5, 1942, 2. PAAA, R 49.165 (Akten betreffend Konferenzen und Verhandlungen [...]).

114 [Auswärtiges Amt (Deutsches Reich)] an das Reichsministerium des Innern (Durchdruck), signed Albrecht, December 30, 1942. PAAA, Berlin, R 49.165 (Akten betreffend Konferenzen und Verhandlungen [...]).

115 Der Reichsminister des Innern (Deutsches Reich) and das Auswärtige Amt, Schnellbrief, signed Weber, January 9, 1943, 1. PAAA, R 49.165 (Akten betreffend Konferenzen und Verhandlungen [...]).

116 Der Reichsminister des Innern (Deutsches Reich) an das Auswärtige Amt, Schnellbrief, signed Weber, January 9, 1943. PAAA, R 49.165 (Akten betreffend Konferenzen und Verhandlungen [...]); citation: 2 (translation: CK).

117 [Auswärtiges Amt (Deutsches Reich)], Aufzeichnung (Internal Note), signed Weyrauch, June 17, 1943, 3. PAAA, R 49.165 (Akten betreffend Konferenzen und Verhandlungen [...]).

118 [Auswärtiges Amt (Deutsches Reich)], Internal Note, written by Weyrauch, no date (attached to document of June 16, 1943). PAAA, R 49.165 (Akten betreffend Konferenzen und Verhandlungen [...]).
119 Translation of title: CK.
120 Der Reichsminister des Innern (Deutsches Reich) an das Auswärtige Amt, signed Müssemeier, May 25, 1943, 1. PAAA, R 49.165 (Akten betreffend Konferenzen und Verhandlungen [...]).
121 Hans Langner, 'Tagung zur Gründung einer zwischenstaatlichen Arbeitsgemeinschaft auf veterinärem Gebiet', *Deutsches Tierärzteblatt. Amtsblatt der Reichstierärztekammer* 10, no. 15/16 (1943): 85–90; Hans Langner, 'Tagung zur Gründung einer zwischenstaatlichen Arbeitsgemeinschaft auf veterinärem Gebiet (Fortsetzung)', *Deutsches Tierärzteblatt. Amtsblatt der Reichstierärztekammer* 10, no. 17/18 (1943): 95–8. See also: Insenhöfer, *Friedrich Weber*, 167–71. A French translation appeared in the Bulletin of the OIE: La Conférence Vétérinaire de Presbourg, in: *OIE Bulletin* 22 (1943), 217–26.
122 [Auswärtiges Amt (Deutsches Reich)], Aufzeichnung (Internal Note), signed Weyrauch, June 17, 1943, 1–2. PAAA, R 49.165 (Akten betreffend Konferenzen und Verhandlungen [...]).
123 Langner, 'Tagung zur Gründung einer zwischenstaatlichen Arbeitsgemeinschaft', 86.
124 [Auswärtiges Amt (Deutsches Reich)], Internal Note, signed Weyrauch, July 1943 (no day), 1. PAAA, R 49.165 (Akten betreffend Konferenzen und Verhandlungen [...]).
125 Protokoll über das Ergebnis der Arbeitstagung der Vertreter der Veterinärverwaltungen des Königreichs Bulgarien, des Deutschen Reichs, des Unabhängigen Staates Kroatien, des Befehlshabers in Serbien, der Slowakischen Republik und des Königreichs Ungarn in der Zeit vom 23. bis 30. Juni 1943. PAAA, R 49.165 (Akten betreffend Konferenzen und Verhandlungen [...]).
126 Protokoll über das Ergebnis der Arbeitstagung [...], Anlage 1. PAAA, R 49.165 (Akten betreffend Konferenzen und Verhandlungen [...]).
127 Langner, 'Tagung zur Gründung einer zwischenstaatlichen Arbeitsgemeinschaft (Fortsetzung)', 96.
128 Langner, 'Tagung zur Gründung einer zwischenstaatlichen Arbeitsgemeinschaft', 86 (translation: CK).
129 Der Reichsminister des Innern (Deutsches Reich) an das Auswärtige Amt (Weyrauch), signed Müssemeier, August 26, 1944. PAAA, R 49.165 (Akten betreffend Konferenzen und Verhandlungen [...]).
130 Réunion Extraordinaire du Comité de l'Office (Juillet 1945), Procès-Verbaux des Séances, in: *OIE Bulletin* 24 (1945), 1–12, here 11.
131 The officers were Dr. Heinz Rievel followed by Dr. Oppermann (first name unknown); both were sons of professors at the Hannover veterinary school (Heinrich Rievel, Theodor Oppermann). OIE: Leclainche à Monsieur le Ministre des Affaires Etrangères (France, Vichy), August 30, 1943. ADLC: Unions, 429QO/335 (Office International des Epizooties, 1940–44); also: Leclainche, in: Compte Rendu de la Quatorzième Session du Comité de l'Office, Procès-Verbaux des Séances, in: *OIE Bulletin* 26 (1946), 191.
132 OIE: Leclainche à Monsieur le Ministre des Affaires Etrangères (France, Vichy), August 30, 1943. ADLC: Unions, 429QO/335 (Office International des Epizooties, 1940–44).
133 Flückiger to Leclainche, May 24, 1943. BAR, 7270(B)/1968/171, Dossier 19/2: OIE, Allgemeine Korrespondenz 1944.
134 Leclainche, in: Réunion Extraordinaire du Comité de l'Office (Juillet 1945), Procès-Verbaux des Séances, in: *OIE Bulletin* 24 (1945), 11.

135 Leclainche to Flückiger, September 26, 1944 (stamp: October 18, 1944). BAR, 7270(B)/1968/171, Dossier 19/2: OIE, Allgemeine Korrespondenz 1944.

136 Réunion Extraordinaire du Comité de l'Office (Juillet 1945), Procès-Verbaux des Séances, in: *OIE Bulletin* 24 (1945), 10–11.

137 On the early history of the FAO: Amy L. S. Staples, *The Birth of Development. How the World Bank, Food and Agriculture Organization, and World Health Organization Changed the World, 1945–1965* (Kent, Ohio: Kent State University Press, 2006).

138 Iris Borowy, 'Maneuvering for Space. International Health Work of the League of Nations during World War II', in *Shifting Boundaries of Public Health*, ed. Susan Gross Solomon, Lion Murard, and Patrick Zylberman (Rochester, NY: University of Rochester Press, 2008), 95. For the dissolution of the OIHP see also: United Nations Treaty Collection, 'Protocol Concerning the Office International d'Hygiène Publique. New York, 22 July 1946', United Nations Treaty Collection, accessed March 19, 2021, http://treaties.un.org/doc/Treaties/1947/10/19471020%2006-23%20AM/Ch_IX_02p.pdf.

139 For the history of UNRRA see: Jessica Reinisch, 'Internationalism in Relief: The Birth (and Death) of UNRRA', *Past and Present* 210, no. Supplement 6 (2011): 258–89.

140 For an overview on activities in public health from the perspective of UNRRA's director of health: Wilbur A. Sawyer, 'Achievements of UNRRA as an International Health Organization', *American Journal of Public Health and the Nation's Health* 37, no. 1 (January 1947): 41–58.

141 Sawyer, 'Achievements of UNRRA', 55.

142 About the UNRRA programme of agricultural and health relief: George Woodbridge, *UNRRA. The History of the United Nations Relief and Rehabilitation Administration*, vol. 1 (New York: Columbia Univ. Press, 1950), 487–92; Jessica Reinisch, 'Introduction: Relief in the Aftermath of War', *Journal of Contemporary History Journal of Contemporary History* 43, no. 3 (2008): 382–5. About UNRRA's projects to support livestock production and its activities against rinderpest see in detail: Amanda Kay McVety, *The Rinderpest Campaigns: A Virus, Its Vaccines, and Global Development in the Twentieth Century* (Cambridge: Cambridge University Press, 2018), 86-120.

143 L'Office International des Epizooties de 1939 à 1946, in: *OIE Bulletin* 26 (1946), 10–11.

144 OIE, Rapport à Monsieur le Président Flückiger sur les Relations établies entre l'UNRRA et l'Office International des Epizooties, signed Leclainche, July 23, 1945, 1. BAR, E 2001 C 1000/1533, BD 492: Office International des Epizooties, Généralités.

145 OIE, Rapport à Monsieur le Président Flückiger sur les Relations établies entre l'UNRRA et l'Office International des Epizooties, signed Leclainche, July 23, 1945, 2. BAR, E 2001 C 1000/1533, BD 492: Office International des Epizooties, Généralités.

146 OIE, Rapport à Monsieur le Président Flückiger sur les Relations établies entre l'UNRRA et l'Office International des Epizooties, signed Leclainche, July 23, 1945, 2. BAR, E 2001 C 1000/1533, BD 492: Office International des Epizooties, Généralités.

147 L'Office International des Epizooties de 1939 à 1946, in: *OIE Bulletin* 26 (1946), 13.

148 L'Office International des Epizooties de 1939 à 1946, in: *OIE Bulletin* 26 (1946), 11.

149 Réunion Extraordinaire du Comité de l'Office (Juillet 1945), Procès-Verbaux des Séances, in: *OIE Bulletin* 24 (1945), 2.

150 Leclainche, in: Réunion Extraordinaire du Comité de l'Office (Juillet 1945), Procès-Verbaux des Séances, in: *OIE Bulletin* 24 (1945), 3 (translation: CK).

151 Agnes Roman, 'UNRRA in China', *Far Eastern Survey* 14, no. 22 (November 7, 1945): 322. In detail about UNRRA's anti-rinderpest activities, especially in China: McVety, *The Rinderpest Campaigns*, 86–120.

152 L'Office International des Epizooties de 1939 à 1946, in: *OIE Bulletin* 26 (1946), 13; La Conférence Vétérinaire de Sydney, in: *OIE Bulletin* 25 (1946), 73–101.

153 L'Office International des Epizooties de 1939 à 1946, in: OIE *Bulletin 26* (1946), 13. About the Southwest Pacific Area Veterinary and Livestock Secretariat and its relations with FAO see: La Conférence de la Santé Animale à Londres, in: *OIE Bulletin* 25 (1946), 165–9, here 168 (d).

154 A summary of the debates appeared in the OIE Bulletin: La Conférence de la Santé Animale à Londres, in: *OIE Bulletin* 25 (1946), 165–9. The conference included national representatives from France, British India, South Africa, Great Britain, USA, as well as from the UNRRA and OIE; the delegates of Brazil, USSR and China could not attend the meeting.

155 Rapport établi par le Professeur Flückiger à l'Intention de l'Office International des Epizooties à Paris sur les Résultats de la Réunion du Sous-Comité pour la Santé Animale de la FAO du 31 Mars au 4 Avril 1947, 2–3. BAR, 7270(B)/1968/171, Dossier 19/2 (1947): Tätigkeitsbericht Seucheamt (sic) 1927–47. For the report on the FAO conference in London published in the OIE Bulletin: Conférence de la Santé Animale à Londres, in: *OIE Bulletin* 25 (1946), 165–9.

156 Conférence de la Santé Animale à Londres, Rapports de la FAO avec les Organisations Vétérinaires Existantes, in: *OIE Bulletin* 25 (1946), here 168 (c); Food and Agricultural Organization of the United Nations: Conférence réunie à Copenhague du 13 au 14 Septembre 1946, in: *OIE Bulletin* 27 (1947), 485–9. The conference reports were written by Flückiger, see: Rapports de l'Office International des Epizooties (OIE) avec la Food Agricultural Organization (FAO), in: *OIE Bulletin* 27 (1947), 482–4, here 483. For Leclainche's report about the meetings to the OIE committee: Compte Rendu de la Quatorzième Session du Comité de l'Office, Procès-Verbaux des Séances, in: *OIE Bulletin* 26 (1946), 195–7.

157 Albania, French Equatorial Africa, French West Africa, Algeria, Austria, Australia, Belgium, Bulgaria, Denmark, Egypt, France, Great-Britain, British India, Luxembourg, Madagascar, Morocco, New Zealand, the Netherlands, Poland, Portugal, Sweden, Switzerland, Czechoslovakia, Tunisia, Turkey, USSR. The delegates of Indo-China, Italy and Hungary were excused. Compte Rendu de la Quatorzième Session du Comité de l'Office, Procès-Verbaux des Séances, in: *OIE Bulletin* 26 (1946), 123–5.

158 Emmanuel Leclainche, 'La Situation Sanitaire de l'Europe en 1946', *OIE Bulletin* 26 (1946): 23 (translation: CK).

159 Flückiger, in: Compte Rendu de la Quatorzième Session du Comité de l'Office, Procès-Verbaux des Séances, in: *OIE Bulletin* 26 (October 1946), 127–8; see also 200–1.

160 Compte Rendu de la Quatorzième Session du Comité de l'Office, Procès-Verbaux des Séances, in: *OIE Bulletin* 26 (October 1946), 221–2.

161 Leclainche, in: Compte Rendu de la Quatorzième Session du Comité de l'Office, Procès-Verbaux des Séances, in: *OIE Bulletin* 26 (October 1946), 195.

162 Food and Agriculture Organization of the United Nations, Standing Advisory Committee on Agriculture: Minutes of Meeting of Subcommittee on Animal Health, Washington D.C., March 31–April 4, 1947. BAR, 7270(B)/1968/171, Dossier 19/2: FAO Unterausschuss für tierische Gesundheitspflege, Konferenz in Washington, 31.3–4.4.1947. For a summary in the OIE Bulletin: Conférence

réunie à Washington du 31 mars au 4 avril 1947. Rapport de M. le Professeur Flückiger, in: *OIE Bulletin* 27 (1947), 489–98. On the meeting of the FAO subcommittee on animal health with regard to discussions on rinderpest see: McVety, *The Rinderpest Campaigns*, 127.

163 The International Office of Epizootics, 1927–47 (author and date not mentioned; presumably Flückiger for the meeting of March 31–April 4 1947), 5–6. BAR, 7270(B)/1968/171, Dossier 19/2: Internationales Tierseuchenamt, Paris, Bericht 1927–47/Tätigkeitsbericht Seucheamt 1927–47, vorgelegt anlässlich der Sitzung des Subkomittees in Washington.

164 The International Office of Epizootics, 1927–47 (author and date not mentioned; presumably Flückiger for the meeting of March 31–April 4 1947), 6. BAR, 7270(B)/1968/171, Dossier 19/2: Internationales Tierseuchenamt, Paris, Bericht 1927–47/Tätigkeitsbericht Seucheamt 1927–47, vorgelegt anlässlich der Sitzung des Subkomittees in Washington.

165 Food and Agriculture Organization of the United Nations, Standing Advisory Committee on Agriculture: Minutes of Meeting of Subcommittee on Animal Health, Washington D.C., March 31–April 4, 1947, 10–11. BAR, 7270(B)/1968/171, Dossier 19/2: FAO Unterausschuss für tierische Gesundheitspflege, Konferenz in Washington, 31.3–4.4.1947.

166 Food and Agriculture Organization of the United Nations, Standing Advisory Committee on Agriculture: Minutes of Meeting of Subcommittee on Animal Health, Washington D.C., March 31–April 4, 1947, 11, 46. BAR, 7270(B)/1968/171, Dossier 19/2: FAO Unterausschuss für tierische Gesundheitspflege, Konferenz in Washington, 31.3–4.4.1947.

167 Compte Rendu de la Quinzième Session du Comité de l'Office, Procès-Verbaux des Séances, in: *OIE Bulletin* 28 (1947), 302–23.

168 Leclainche, in: Compte Rendu de la Quinzième Session du Comité de l'Office, Procès-Verbaux des Séances, in: *OIE Bulletin* 28 (1947), 320.

169 Leclainche, in: Compte Rendu de la Quinzième Session du Comité de l'Office, Procès-Verbaux des Séances, in: *OIE Bulletin* 28 (1947), 314.

170 Projet d'Accord entre l'Organisation des Nations Unies pour l'Alimenation et l'Agriculture et l'Office International des Epizooties, in: Compte Rendu de la Quinzième Session du Comité de l'Office, Procès-Verbaux des Séances, in: *OIE Bulletin* 28 (1947), 347–9.

171 Création des Offices Régionaux des Epizooties, in: Compte Rendu de la Quinzième Session du Comité de l'Office, Procès-Verbaux des Séances, in: *OIE Bulletin* 28 (1947), 357–61.

172 Texte de l'Accord Intervenu entre la Food Agricultural Organization et l'Office International des Epizooties (texte anglais original), September 12, 1947, in: *OIE Bulletin* 27, 499–502, citation: 500.

173 Texte de l'Accord Intervenu entre la Food Agricultural Organization et l'Office International des Epizooties (texte anglais original), September 12, 1947, in: *OIE Bulletin* 27, 499–502, here 500.

174 Texte de l'Accord Intervenu entre la Food Agricultural Organization et l'Office International des Epizooties (texte anglais original), September 12, 1947, in: *OIE Bulletin* 27, 499–502, here 500–1, citation: 501.

175 Texte de l'Accord Intervenu entre la Food Agricultural Organization et l'Office International des Epizooties (texte anglais original), September 12, 1947, in: *OIE Bulletin* 27, 499–502, here 501–2.

176 Résolutions, in: Compte Rendu de la Quinzième Session du Comité de l'Office, Procès-Verbaux des Séances, in: *OIE Bulletin* 28 (1947), 389. See also: McVety, *The Rinderpest Campaigns*, 131.

177 See in detail for the FAO rinderpest campaign: McVety, *The Rinderpest Campaigns*, esp. ch. 4.

178 Résolutions, in: Compte Rendu de la Quatorzième Session du Comité de l'Office, Procès-Verbaux des Séances, in: *OIE Bulletin* 26 (October 1946), 221–2 (citation: 222).
179 La Conférence de Berne pour la Préparation du Vaccin Anti-Aphteux, November 20–22, 1946, in: *OIE Bulletin* 25 (1946), 219–20.
180 Arrangement concernant la Préparation et la Distribution du Vaccine Anti-Aphteux, November 22, 1946, in: *OIE Bulletin* 25 (1946), 220–2.
181 Résolutions, in: Procès-Verbal de la Conférence à Berne par l'Office International des Epizooties pour l'Etude de l'Uniformisation des Méthodes de Préparation du Vaccin Anti-Aphteux (du Septembre 30 au Octobre 3, 1947), in: *OIE Bulletin* 27 (1947), 529–30.
182 Flückiger, 'Internationale Bekämpfung tierischer Seuchen', 21.
183 Flückiger, 'Internationale Bekämpfung tierischer Seuchen', 22.
184 Flückiger, 'Internationale Bekämpfung tierischer Seuchen', 21–2.
185 Flückiger, 'Internationale Bekämpfung tierischer Seuchen', 22–4.
186 Food and Agriculture Organization of the United Nations, 'Reports of the Conference of FAO, Seventh Session, Rome November 23–December 11, 1953, Appendix: Agreement between FAO and OIE', accessed March 19, 2021, http://www.fao.org/docrep/x5576E/x5576e0h.htm#c.%20agreement%20between%20fao%20and%20oie. See also: Flückiger, 'Internationale bekämpfung tierischer Seuchen', 25.
187 World Organisation for Animal Health (OIE), 'History', accessed December 28, 2021, https://www.oie.int/en/who-we-are/mission/history/.
188 For the new FAO/OIE programme on foot-and-mouth disease see: World Organisation for Animal Health (OIE), 'FAO/OIE Global Conference on Foot and Mouth Disease Control', accessed March 19, 2021, http://www.oie.int/eng/A_FMD2012/background.html.

7 Conclusion

After the Second World War, the situation of international efforts of animal disease control had changed considerably in comparison to a quarter of a century earlier. As an urgent problem area cutting across borders, contagious animal diseases were now permanently placed on the agenda of several international organisations, most notably the FAO and the OIE, later joined by the WHO and further governmental and non-governmental organisations. In the following decades, they initiated large-scale projects on animal health by linking the international, national and local levels of anti-disease efforts. The focus was on organising joint international activities, especially with vaccination campaigns, to protect the livelihood of people dependent on animal husbandry, and increasingly on the control of zoonotic diseases. The international projects focusing on animal disease control in the decades before the Second World War had created important fundaments for these initiatives. Since the 1860s, a number of veterinary experts, the main protagonists of these early international initiatives, had tried to attract the attention of politicians and state administrations to the problem area to make it part of international governmental cooperation. Through their activities they made sure that the issue of the containment of animal diseases at the international level was maintained and enhanced over the decades despite political and economic crises. Most importantly, they developed international strategies with a range of concepts and practices to address the cross-border spread of contagious animal diseases, which in many respects formed the basis for future anti-disease activities. They worked to persuade governments to establish permanent international institutions and infrastructures, as well as to provide funding for measures of animal disease control. They also began to conceptualise the idea of international animal vaccination campaigns. Central for the future efforts against animal diseases was the development of a system of international information transfer: veterinary experts active at international platforms aimed to improve the control of animal diseases in the long term and in a sustainable manner by creating a continuously updated, worldwide system of information exchange and transfer, concentrating,

DOI: 10.4324/9781003033998-7

preparing and distributing up-to-date knowledge on the global animal disease situation. Many of the envisaged projects addressing the problems of cross-border animal diseases between the late 19th and the mid-20th centuries were not implemented in practice, but from today's perspective it would not be helpful to dismiss them easily as failures. To address globally relevant problems such as the spread of contagious animal diseases, with a complex set of uncertainties and no clear solutions in sight, simple and linear answers were certainly not expedient. By examining the efforts of historical actors to initiate cross-border activities, discuss practices and international norm-building, and negotiate the advancement of their projects with governments and international organisations, a number of challenges of developing responses to pressing complex problem areas with a global dimension become apparent.

Responses to animal diseases beyond borders

The problems of international efforts to control animal diseases resulted from the multiple uncertainties directly or indirectly accompanying animal diseases, as well as from the political, economic and social contexts of the time which were reflected in the activities of veterinary and scientific internationalism. The complex origins, different modes of transmission and the often unpredictable spatial and temporal dimension of disease spread posed multiple challenges for historical actors. The international political and economic tensions as well as structural asymmetries of power, present on all international platforms, did not make the search for responses and activities in a global context any easier. A central issue around which discussions at international meetings revolved was the question of organising sanitary borders as measures against the spread of pathogens, and the problems presented by the associated practices to reduce animal diseases. Borders took on different forms and functions in the context of animal disease containment. Animal diseases that spread across geographical and political borders undermined the function of state borders and thus increased the pressure on state sovereignty. In countries where animal disease control was the responsibility of government authorities, one of the first and most important responses of governments to emerging animal disease challenges was usually to protect the territory by upgrading state borderlines with sanitary control stations to regulate the import and transit of animals or animal products. Veterinary sanitary border practices included control and documentation, as well as hygiene and physical distancing measures for animals, such as quarantines as a temporal and spatial instrument to control the spread of diseases in a defined area, taking into account the incubation period and the time of contagiousness. Borders in this context functioned as potentially mobile border stations and lines, materially constructed and dynamically reshaped, and as processes of creating demarcations, categorisations, inclusions and exclusions. However, the problem remained that borders could not always stop the spread of a contagious animal disease.

The territorial approach to preventing animal diseases is nevertheless still one of the most important conceptual frameworks of animal disease control today. However, as territorially oriented veterinary sanitary practices showed deficits, especially in global contexts, further and more comprehensive transboundary measures needed to be developed.

At international forums, historical actors—veterinary experts, physicians, politicians, diplomats and civil servants—usually agreed that the existing system of sanitary borderlines as protective barriers against animal diseases showed significant shortcomings, especially in the global context of the late 19th and early 20th centuries. They put forward that, given the situation of many differently organised national and colonial veterinary services, veterinary internationalism needed to address not only the improvement of sanitary border controls, but also to organise the crossing of borders by transports of animals and animal products, as well as initiate activities of prevention. They reflected on the need to connect the categories of 'inside' and 'outside' of territorially organised sanitary systems with cross-border concepts of addressing diseases through both sanitary measures and prophylaxis, especially vaccination, combined with new instruments of international information and cooperative research activities. Still, many suggestions of fighting animal diseases at the international level were not implemented in practice, or were done so only decades later. Especially the projects of intergovernmental agreements to collaborate against the spread of animal diseases proved to be difficult to be realised, with only a limited number of governments actively participating. Similar to other fields of international cooperation, it was an immense problem to advance activities effective on a global scale if the 'great powers' did not show much commitment—even though their livestock sectors were severely affected by cross-border animal disease problems, and many countries were highly dependent on livestock importations. Already in the 19th century, veterinary experts started initiatives to organise common approaches to cooperation in disease control beyond the governmental level, in particular through activities of networking and information transfer within the scientific community. The overlap between the official sphere of government representation and the professional exchange of the scientific community of veterinarians clearly contributed significantly to advancing the efforts of veterinary internationalism.

'Organising trust'

From the middle of the 19th century onwards, participants in international projects discussed the question of how measures to contain the cross-border spread of animal diseases could be effectively implemented. Many experts suggested making new national veterinary regulations as uniform as possible in order to create from the start mutually trustworthy veterinary systems. This should allow a safer import and transit of animals. As many states had only rudimentary or no veterinary services at the end of the 19th century, it seemed an ideal time to introduce a common—and therefore mutually trustworthy—structural solution. The idea of establishing uniform national regulatory systems was taken up in different projects aiming to establish international principles, norms and standards for veterinary regulation. The focus on international organisations as platforms for 'organising trust' offers a new perspective on how historical actors, in this case veterinarians, used and influenced internationalism to foster collaboration on animal disease problems. 'Organising trust' as an instrument of internationalism became especially relevant because it was difficult to find concrete and implementable solutions to problems of animal diseases in a global context. The term refers to the development of communication structures in response to the spread of animal diseases, and to how historical actors used a range of different options and best practices to address animal disease problems. Processes of 'organising trust' formed the basis for international agreements, but were also important when these agreements did not work in practice, were not effective enough, did not have enough cooperation partners or needed additional impetus. In the case of veterinary internationalism, historical actors connected the development of trustful relationships above all with what was considered trustworthy information on the situation of animal diseases. If uniform state sanitary services could not be realised, an effective and centralised international system for the transfer of information on animal diseases would, in the perception of many experts, help to reduce uncertainties about the animal disease situation at a global level, but also to identify when certain scientific standards were not being met. This idea included the collection and dissemination of information through an international bulletin and an international telegraphic service for flexible short-term communication. At another level, it involved the principle of transparent documentation and registration of individual animals and animal products intended for transport through an international health or production certificate. In this perspective, the international platforms served as hubs where the negotiations on 'organising trust' and the discussions on its implementation converged, and at the same time they took on the central role of 'organising' through their work with information as the essential component of trustful relationships. Institutions such as the OIE had no authority to intervene directly in national animal disease policies; they usually only had limited financial resources and no possibility to admonish member states that did not comply with the guidelines or requirements. They promoted trustful relations to work towards multiplying the cooperation partners they needed to produce information on the global situation of animal disease risks. 'Organising trust' was also associated with direct and indirect control: the individual members of the international expert community recognised each other as trustworthy partners if they showed the appropriate academic-scientific background. All elements of 'organising trust' included the idea of an alignment with European standards of veterinary principles as the basis of trust-building measures. Trust as the basis of international cooperation in the field of animal diseases included hierarchies, and the building of trust was based on the asymmetries of power within cooperation partners.

The years after the First World War saw changing structural conditions for veterinary internationalism which resulted in new opportunities for activities of international cooperation. For the first time, there existed international institutions that included animal diseases and the problem of their transboundary dimension into their agenda. Furthermore, new thematic aspects in international work became more prominent in the discussions. Next to the problems of epizootics cutting across borders in context of trade relations, veterinary questions with regard to public health came into the focus of international approaches. In the case of anthrax, a zoonotic disease appearing on a worldwide scale, a joint ILO/LNHO project focused on finding practical ways of local disease prevention to intercept the transmission of pathogens in transport and in the industrial processing of animal products. A LNHO scientific project on rabies tried to advance secure methods of vaccination through a long-term statistical investigation in global contexts. Three international veterinary conventions of the League of Nations focused on the facilitation of international trade in animals and animal products, taking up the debates to contain animal diseases in context of trade since the 19th century. As they were only implemented by a limited number of governments, the conventions did not achieve the expected significance, but were nevertheless an important orientation for future veterinary legislation. Due to the difficulties in creating international arrangements, subsequent veterinary projects of the League of Nations again focused on the idea of producing recommendations and non-binding rules or guidelines. This reversed the idea of 'organising trust': guarantees provided by international conventions did not exist in these cases, but precisely because of the non-binding nature of the guidelines, there was hope that national veterinary authorities and local practitioners would use them as orientation and thus work to improve the respective veterinary systems and control activities. During the same period, the OIE established a new, permanent system of information transfer with the aim to create a culture of transparency on the global animal disease situation. This task proved to be extremely difficult. Many member governments did not or only partially contributed, either because they feared problems with trade or because they simply did not have at their disposal a veterinary system with the necessary scientific criteria. The OIE tried to provide as much information as possible on the global status of animal diseases, while at the same time offering a regular platform for scientific and veterinary exchange. The staff steered the office through the international political tensions of the time by emphasising the technical and non-political nature of its work. In the late 1930s the OIE experts started developing new steps intending to extend international cooperation and networking towards joint international activities, especially with the idea of organising a project of transboundary distribution of vaccines. The idea of closely linking the international with the national and local level to ensure universal vaccination across national borders, in this case against foot-and-mouth disease, for the benefit of animal health in all participating countries regardless of political tensions, was an important and promising initiative. It would have strengthened the OIE's position as international coordinator and underlined its role towards a more active international intervention in the field of animal disease prevention. However, it was not launched because of the war. The National Socialist government tried to use the ideas of international animal disease control for its war economy by starting similar initiatives in its sphere of influence. Under pressure from various directions, the OIE staff had to rely on their international scientific networks and professional authority to continue the office after the Second World War.

Veterinary Eurocentrism in a global context

The power relations and hierarchies at the international meetings on veterinary cooperation shaped the thematic priorities and the structural organisation of the international projects to control animal diseases. The main group of historical actors in the field of animal diseases, the veterinary experts, used the international platforms to develop their professional community, to network with each other, to explore their scope of action and to advance their veterinary projects. They were the ones who, in most cases, decided on the agenda of the international meetings on animal diseases and pushed forward new projects. The idea that experts should work out politically neutral solutions to complex so-called technical problem areas was influential. It served as a legitimation to continue international cooperation and dialogue despite political, ideological and economic tensions, and it underlined the professional authority of veterinary experts on international platforms. Although they had in most cases a similar scientific and veterinary background, the participants of international meetings did not form a homogenous group. They were a community that—in part—pursued certain common interests and goals, while many of them at the same time also represented the interests of their governments, which were not always congruent. In different ways, the experts were able to influence their respective national veterinary policies. However, they often expressed disappointment with the reluctant and hesitant attitudes of governments and official decision makers. In the field of animal diseases, governmental internationalism was indeed slow to find concrete and binding solutions. Many governments preferred veterinary approaches in a national or imperial framework to a direct commitment to international conventions, or they concluded bilateral agreements with trading partners and ignored both the dangerous global dimension of the problem area and the activities on international platforms. The most important long-term success for experts in the period before the Second World War was therefore certainly the establishment of the OIE, created by an international agreement in 1924. But the OIE had no authority to intervene in national veterinary policy. Veterinary experts were therefore faced with a constant dilemma when working on international platforms: the measures they discussed in detail to contain animal diseases ultimately depended on the official implementation and legitimation through intergovernmental agreements. They responded to this problem by expanding their scientific networks and focusing on the production of non-binding recommendations with the aim of influencing national policies more indirectly but on a broader scale. The influence of the experts therefore depended on the respective context and situation: new possibilities of activity, also on the international level, resulted from their professional authority, involvement in international platforms and the network of official and informal connections, but at the same time, in most cases a structural limitation of influence remained clear. The weight of individual experts' voices within their scientific community in the international context varied greatly depending on their country of origin, scientific reputation and professional position. With their dominant position, the experts from the Western and Central European countries not only influenced what was considered internationally relevant, but also determined the structural conditions and norm-setting of veterinary internationalism. One of the most important hierarchical categories continued to be the idea of the 'civilised' state, which, although rarely mentioned, was reflected in the discussions by views on the quality and structure of veterinary services. At the international conferences of the late 19th century, veterinary researchers and authorities from Western and Central European countries routinely held Russia responsible for the spread of rinderpest epizootics into Europe. After the First World War, several hierarchies of veterinary internationalism became manifest at the International Conference for the Study of Epizootics in 1921 in Paris. Firstly, the voices of veterinary experts from smaller, mostly Southern European countries appeared only marginally in the minutes of the meetings; the debates were clearly dominated by representatives of Western and Central European countries. Most initiatives were based on suggestions from French, German, Dutch, Belgian, Italian or British experts. In the case of the creation of the OIE, the commitment of the French government was instrumental. As a result, French experts and their scientific networks often played a significant role in most international veterinary debates until the 1940s. Secondly, the conference of 1921 aimed to establish a new and global approach to animal disease problems, and at the same time, the debates revealed the constructed gap between Europe and the rest of the world. Dantès Bellegarde, participating as the delegate from Haiti, commented directly on the point that 'international' practices seemed to imply 'European', and this approach had consequences.[1] There was dissatisfaction with the one-sided direction of veterinary internationalism and the universal-global claim of European approaches to animal disease control, and with the resulting omissions, blind spots and stagnations, even though such voices often were the quiet ones. 'European', with Europe as a constructed region, was not congruent with 'Western' in these discourses; the United States and Canada, for example, were only marginal participants in intergovernmental veterinary work until 1946, and North

American experts for long decades did not play a significant role on any of the international platforms analysed.

The discrepancy between European-focused responses to animal disease problems and the globality of their spread remained a feature of international animal disease activity until the middle of the 20th century. In the 1860s, 1870s and 1920s it was the arrival of 'foreign' rinderpest in Western Europe that motivated new international initiatives. The globality of animal diseases was always considered an important element of what made them potentially dangerous, and their global character was further emphasised when scientific research revealed more details about specific pathogens and their spread. Veterinary experts from Western and Central Europe interpreted the spread of animal diseases as a threatening consequence of a new globalised modernity, and they used the global dimension of animal diseases both to stimulate discussion and to legitimise their activities. In the case of the OIE, the responsible veterinary experts tried to approach the global dimension of the spread of animal diseases by integrating as many member states as possible in order to establish a comprehensive worldwide information and communication network. The office aimed to include worldwide data on the disease situation in its statistics and publications. Indeed, in order to follow the original idea of an effective exchange of information to prevent the spread of animal diseases and to create trustful relationships for trading partners, the office needed the regular contacts with non-European veterinary administrations and experts from independent states as well as from colonies. However, many countries lacked the necessary veterinary infrastructure and financial resources to participate effectively in international projects. This fundamental problem was frequently mentioned but not adequately addressed for decades. Imperial powers also built up their own imperial veterinary infrastructure and networks. The problem that European activities exported animal diseases to other parts of the world through colonialism, trade and migration was usually only a marginal topic of discussion on international platforms, even though some of these disease transfers were devastating in a global perspective. The tension between the global context of disease transmission and the European focus of international work against animal diseases became even more pronounced after the Second World War. Both world wars led to enormous movements of people, animals and animal products on a global scale, and they brought about profound ecological and environmental changes in many parts of the world. After the First World War, the need to contain the spread of animal diseases in global contexts led to the creation of a new institution in the form of the OIE. After the Second World War, the OIE faced the major problem that its institutional structure and funding was too limited to address issues of animal health in general in a broader global context. International cooperation in the fight against animal diseases needed to reorganise itself in the post-war years. The focus on the protection of Europe, together with the Eurocentrism of international veterinary concepts, practices and institutions, had become problematic with regard to the global situation of animal health, and for Europe's animal disease situation as well.

The complex problem area of animal diseases always needed answers on multiple levels, as the discussions at the international platforms showed, especially with view to their global dimension. The global context of animal diseases refers, beyond the actual spatial spread of pathogens, also to the multifaceted uncertainties associated with the spread of contagious diseases. Thus, the danger of an epidemic or an epizootic could exist even if it did not appear in a certain place. Another dimension of the globality of animal diseases relates to the exploitation of people, animals and resources in conjunction with environmental changes. Animal diseases were regarded as an economic problem, and increasingly as a public health concern as well. Veterinary experts on international platforms rarely discussed the political, economic and societal contexts of their emergence and global spread, nor the role of colonialism and industrialisation. However, the history of animal diseases and their containment is closely related to global economic relations, the modernisation of agriculture and environmental transformations. The history of animal husbandry, agriculture, industrial processing of animals and animal products, wildlife hunting, and their respective environmental and ecological consequences is inevitably connected to histories of capitalism. The transport of huge numbers of animals over great distances was linked to economic profit, colonial aspirations, warfare or all at the same time. When it came to combating animal diseases, the European societies that dominated in the international discussions clearly prioritised their problems over the concerns of the non-European world, despite their responsibilities. The history of animal diseases sheds light on the entangled circulation of animals, pathogens, money, ideas and concepts; on the role of global agribusiness and industrial mass livestock farming in the spread of diseases; and on social, geographical and ecological imbalances and inequalities. The question of power asymmetries and responsibilities in a global context is therefore also reflected in the history of animal disease problems in global contexts.

Note

1 Dantès Bellegarde, in: République Française, Ministère de l'Agriculture, ed., *Conférence Internationale pour l'Etude des Epizooties. Paris (25–28 Mai 1921). Compte Rendu Sténographique* (Paris: Imprimerie Nationale, 1921), 97.

Brief glossary of cited animal diseases

Anthrax (French: charbon, maladie du charbon, fièvre charbonneuse; German also: Milzbrand): acute infectious and highly contagious bacterial disease, which can be transmitted to humans through the bacterial spores.

Bovine pleuropneumonia (also: contagious bovine pleuropneumonia; French: pleuropneumonie contagieuse bovine; German: Lungenseuche der Rinder): highly contagious chronic bacterfial infection of cattle, buffalos and related animals, with the symptom of lung inflammation.

Bovine spongiform encephalopathy (BSE, also informally: mad cow disease; French: encéphalopathie spongiforme bovine; German: Bovine spongiforme Enzephalopathie): disease of the nervous system of cattle, caused by abnormal proteins (prions) and thought to be transmissible to humans.

Bovine tuberculosis (French: tuberculose bovine; German: Tuberkulose der Rinder, Rindertuberkulose): bacterial disease of animals (livestock and wild animals) which can be transmitted to humans via droplet infection or raw milk.

Brucellosis (French: brucellose; German: Brucellose): contagious bacterial disease of livestock; as a zoonotic disease it can be transmitted to humans, for example via raw milk.

Coronavirus disease 2019 (COVID-19): highly contagious infectious disease in humans caused by the SARS-CoV-2 virus; thought to be of animal origin, with emergence in different animal species.

Dourine (also: covering sickness; French: dourine; German: Beschälseuche): parasitic (chronic or contagious) disease of horses, donkeys and mules transmitted during sexual intercourse.

Foot-and-mouth disease (French: fièvre aphteuse; German: Maul- und Klauenseuche): viral disease of different cloven-hoofed livestock species, with often severe economic effects because of its high contagiousness.

Fowl cholera (also: chicken cholera, avian cholera; French: choléra des poules; German: Geflügelcholera): bacterial infectious disease of poultry and wild birds.

Glanders (French: morve, German: Rotz): infectious zoonotic bacterial disease of horses, donkeys and mules; it can be transmitted to numerous other species, including wild animals, as well as to humans.

Influenza (French: grippe; German: Virusgrippe): infectious contagious viral disease; with different viral subtypes affecting animals and with a zoonotic potential (avian influenza, swine influenza).

Mange (French: gale animale; German: Räude): parasitic skin disease (mite infestation) of livestock and wild animals with a zoonotic potential; scabies in humans is caused by a different mite infestation.

Rabies (French: rage; German: Tollwut, Wut, Lyssa): zoonotic viral disease infecting animals (mammals) and humans, usually transmitted through contact with another animal, mostly dogs.

Rinderpest (also: cattle plague, steppe murrain; French: peste bovine; German: Rinderpest): highly contagious infectious viral disease of cattle and other cloven-hoofed animals; declared eradicated in 2011.

Sheep pox (French: clavelée; German: Pockenseuche der Schafe und Ziegen): viral disease of sheep and goats.

Swine fever (also: classical swine fever, hog cholera; French: peste porcine; German: Schweinepest): highly contagious viral disease of pigs, frequently transmitted via wild pigs or vectors such as food or waste.

Trichinosis (also: trichinellosis; French: trichinose, trichinellose; German: Trichinose, Trichinellose, Trichinenkrankheit): infectious disease caused by a parasitic roundworm in pigs and other mammals with a zoonotic dimension; to humans, it can be transmitted, for example, via raw meat.

See for detailed information: World Organisation for Animal Health (OIE), 'Animal Diseases: Portal (OIE List)', accessed 30 May 2021, https://www.oie.int/en/what-we-do/animal-health-and-welfare/animal-diseases/.

Archival sources and bibliography

Archival sources

Archives Diplomatiques/Diplomatic Archives, France, La Courneuve (ADLC)

Unions Internationales, Période 1907–1944, 429QO/332.
Unions Internationales, Période 1907–1944, 429QO/333.
Unions Internationales, Période 1907–1944, 429QO/334.
Unions Internationales, Période 1907–1944, 429QO/335.
Unions Internationales, Période 1907–1944, 429QO/357.
Unions Internationales, Période 1907–1944, 429QO/498.
Unions Internationales, Période 1907–1944, 429QO/658.

Hoover Institution Archives, Stanford University, Stanford, CA, USA (Hoover Institution Archives)

Germany, DKZ, Box 190: Internationaler Landwirtschaftskongress 1939.
Germany, DKZ, Box 307: Tierärzte-Kongress 1863–1942; Tierseuchen – Internationales Büro zur Bekämpfung.

International Labour Organization: International Labour Office, Historical Archives, Geneva (ILO Archives)

Hy 550/8.
D751/100/5.

League of Nations Archives, Geneva (LON)

R902: 12B/30055/9891; 12B/30700/30700; 12B/32307/30700; 12B/34398/30700; 12B/40053/30700; 12B/43801/30700.
R948: 12B/41192/41192; 12B/50886/41192; 12B/52551/41192.
R2774: 10D/2049/272; 10D/272/272.
R2780: 10D/1275/309; 10D/1477/309.
R2795: 10D/16979/1660.
R4373: 10A/38438/673.
R4379: 10A/33914/1513.

R4427: 10A/15577/15577; 10A/33868/15577.
R4432: 10A/19997/19997.
R4460: 10A/36175/33950.
R4467: 10A/37646/34956.
R4468: 10A/37431/37431.
R5858: 8A/3617/1214.
R5880: 8A/12251/3856.
R6066: 8A/1983/1167.
R6068: 8A/18256/11384; 8A/43353/1167.
R6110: 8A/34138/11384; 8A/11384/11384; 8A/18256/11384 /8A/43055/11384.

Politisches Archiv des Auswärtigen Amts/Political Archive of the Federal Foreign Office, Berlin, Germany (PAAA).

R 49.159.
R 49.165.
Paris 1.376 (Botschaft Paris: 246 Veterinärwesen).

Schweizerisches Bundesarchiv/Swiss Federal Archives, Berne, Switzerland (BAR)

7270(B)/1968/171, Dossier 19/2.
E 2001 C 1000/1533, BD 492.

Bibliography

Periodicals

Office International des Epizooties (OIE): *Bulletin de l'Office International des Epizooties* (OIE Bulletin), vol. 1 (1927/28)-28 (1947).

Deutsche Vierteljahrsschrift für öffentliche Gesundheitspflege. Organ des Deutschen Vereins für öffentliche Gesundheitspflege, 1 (1869)–47 (1915), *incl. Jahresbericht über die Fortschritte und Leistungen auf dem Gebiet der Hygiene.*

Printed books and articles

'A Word to Belgium'. *The Times*, 29 May 1919.

Academia Nacional de Agronomia y Veterinaria. 'Lignières, José,' n.d. Accessed 19 March 2021. http://www.anav.org.ar/index.php?option=com_content&view=article&id=190:lignieres-jose-med-vet&catid=43:de-numero-fallecidos&Itemid=69.

'Agriculture.gouv.fr: Les Ministres de l'Agriculture – IIIe République (1870–1940). Joseph-Honoré Ricard,' n.d. Accessed 19 March 2020. http://agriculture.gouv.fr/220-Joseph-Honore-RICARD.

Akami, Tomoko. 'A Quest to Be Global: The League of Nations Health Organization and Inter-Colonial Regional Governing Agendas of the Far Eastern Association of Tropical Medicine 1910–25'. *The International History Review* 38, no. 1 (2016): 1–23.

Alnot, Laure, and Guilhem Pistre. 'La Vie et l'Oeuvre d'Emmanuel Leclainche (1861–1953)'. *Bulletin de la Société Française d'Histoire de la Médecine et des Sciences Vétérinaires* 4, no. 1 (2005): 94–103.

Andrews, John B. *Anthrax as an Occupational Disease*. Bulletin of the United States Bureau of Labor Statistics 205. Washington, DC: Government Printing Office, 1917.

Arloing, Saturnin, ed. *Ve Congrès International de Médecine Vétérinaire, tenu à Paris du 2 au 8 Septembre 1889: Recueil de Rapports*. Paris: Asselin et Houzeau, 1889.

Asdal, Kristin, Steve Hinchliffe, and Tone Druglitrø, eds. *Humans, Animals and Biopolitics: The More-than-Human Condition*. Abingdon and New York: Routledge, 2017.

Baberowski, Jörg, ed. *Was ist Vertrauen? Ein interdisziplinäres Gespräch*. Frankfurt am Main and New York: Campus Verlag, 2014.

Baldwin, Peter. *Contagion and the State in Europe, 1830–1930*. Cambridge and New York: Cambridge University Press, 1999.

Balinska, Marta Aleksandra. *For the Good of Humanity: Ludwik Rajchman, Medical Statesman*. New York: Central European University Press, 1998.

Barrett, Thomas, Paul-Pierre Pastoret, William P. Taylor, Gordon Scott, and A. Provost. 'Rinderpest – An Old and Worldwide Story: History to c. 1902'. In *Rinderpest and Peste des Petits Ruminants. Virus Plagues of Large and Small Ruminants*, edited by Thomas Barrett, Paul-Pierre Pastoret, and William P. Taylor, 86–104. London and Burlington, MA: Elsevier/Academic, 2006.

Barth, R., H. L. Bock, and G. Korger. 'Geschichte der Tollwutschutzimpfung des Menschen im deutschsprachigen Raum'. *Medizinhistorisches Journal* 24, no. 3–4 (1989): 333–51.

Bashford, Alison, ed. *Medicine at the Border: Disease, Globalization and Security, 1850 to the Present*. Houndmills, Basingstoke: Palgrave Macmillan, 2006.

———. '"The Age of Universal Contagion": History, Disease, and Globalization'. In *Medicine at the Border. Disease, Globalization and Security, 1850 to the Present*, 1–17. Houndmills, Basingstoke: Palgrave Macmillan, 2006.

Bauerfeind, R., P. Kimmig, H. G. Schiefer, T. Schwarz, W. Slenczka, and H. Zahner. *Zoonosen: Von Tier zu Mensch übertragbare Infektionskrankheiten*. Köln: Deutscher Ärzteverlag, 2013.

Bazin, Hervé. 'L'Histoire des Vaccinations. 1ère Partie: De la Variolisation à la Vaccination'. *Bulletin de la Société Française d'Histoire de la Médecine et des Sciences Vétérinaires* 12 (2012): 133–49.

Berger, Silvia. *Bakterien in Krieg und Frieden. Eine Geschichte der medizinischen Bakteriologie in Deutschland 1890–1933*. Göttingen: Wallstein, 2009.

Bergés, Pedro. *Instituto Internacional de la Fiebre Aftosa y Oficina Sanitaria Internacional. Trabajos realizados para su Creación en Buenos Aires (1920–1924)*. Buenos Aires: Revista de Industria Lechera y Ganaderia, 1924.

Bisping, Walter. 'Zur Geschichte der Bekämpfung der Rindertuberkulose'. *Berliner und Münchener Tierärztliche Wochenschrift* 111, no. 10 (1998): 362–67.

Blancou, Jean. *Histoire de la Surveillance et du Contrôle des Maladies Animales Transmissibles*. Paris: OIE, 2000.

Blancou, Jean, and Claude Meurier. 'Brève Histoire de Quelques Laboratoires Nationaux Vétérinaires Français: Première Partie'. *Bulletin de la Société Française d'Histoire de la Médecine et des Sciences Vétérinaires* 6 (2006): 79–107.

Borgards, Roland, ed., *Tiere. Kulturwissenschaftliches Handbuch*. Stuttgart: Metzler, 2016.

Borowy, Iris. *Coming to Terms with World Health. The League of Nations Health Organisation, 1921–1946*. Frankfurt am Main and New York: Peter Lang, 2009.

———. 'Maneuvering for Space. International Health Work of the League of Nations during World War II'. In *Shifting Boundaries of Public Health*, edited by Susan Gross Solomon, Lion Murard, and Patrick Zylberman, 87–113. Rochester, NY: University of Rochester Press, 2008.

Borowy, Iris, and Wolf D. Gruner, eds. *Facing Illness in Troubled Times: Health in Europe in the Interwar Years, 1918–1939*. Frankfurt am Main and New York: Peter Lang, 2005.

Bossneck, Joachim. 'Eduard von Hering'. In *Neue Deutsche Biographie*, 8:617. Berlin: Duncker & Humblot, 1969.

Boudreau, Frank G. *Ancient Diseases – Modern Defences. The Work of the Health Organization of the League of Nations*. New York: Columbia University Press, 1939.

———. 'International Coöperation in Hygiene'. *American Journal of Public Health* 27, no. 11 (1937): 1102–1107.

Brantz, Dorothee, ed. *Beastly Natures: Animals, Humans, and the Study of History*. Charlottesville: University of Virginia Press, 2010.

———. 'How Parasites Make History: On Pork and People in Nineteenth-Century Germany and the United States'. *Bulletin of the GHI Washington*, no. 36 (2005): 69–79.

———. 'Slaughter in the City. The Establishment of Public Abattoirs in Paris and Berlin, 1780–1914'. PhD Thesis, University of Chicago, 2003.

Bresalier, Michael, Angela Cassidy, and Abigail Woods. 'One Health in History'. In *One Health. The Theory and Practice of Integrated Health Approaches*, edited by Jakob Zinsstag, Esther Schelling, David Waltner-Toews, Maxine Whittaker, and Marcel Tanner, 1–15. Wallingford, Boston: Cabi, 2015.

Briel, Cornelia, and Hans Erich Bödeker. *Beschlagnahmt, erpresst, erbeutet: NS-Raubgut, Reichstauschstelle und Preußische Staatsbibliothek zwischen 1933 und 1945*. Berlin: Akademie Verlag, 2013.

Brown, Fred. 'The History of Research in Foot-and-Mouth Disease'. *Virus Research* 91, no. 1 (2003): 3–7.

Brown, Karen. 'Tropical Medicine and Animal Diseases: Onderstepoort and the Development of Veterinary Science in South Africa, 1908–1950'. *Journal of Southern African Studies* 31, no. 3 (2005): 513–29.

Brown, Karen, and Daniel Gilfoyle, eds. *Healing the Herds. Disease, Livestock Economies, and the Globalization of Veterinary Medicine*. Athens, Ohio: Ohio University Press, 2010.

Brumme, Martin Fritz. 'Lebenswege eines wilhelminischen Bürgers. Biographische und anatomiegeschichtliche Untersuchungen zu Reinhold Schmaltz (1860–1945)'. PhD Thesis, Freie Universität Berlin, 1992.

Bulling, Eberhard, and Arno Schönberg. 'Robert von Ostertag (1864–1940): A Veterinarian Contemporary with R. Virchow and R. Koch'. *Historia Medicinae Veterinariae* 24, no. 4 (1999): 97–119.

Bürgi, Markus. 'Zangger, Rudolf'. *Historisches Lexikon der Schweiz*, n.d. Accessed 1 April 2021. https://hls-dhs-dss.ch/de/articles/003731/2014-02-03/.

Buszko, J. 'Julian Nowak'. In *Österreichisches Biographisches Lexikon 1815–1950*, 7:165. Wien: Verlag der Österreichischen Akademie der Wissenschaften, 1978.

Castonguay, Stéphane. 'Creating an Agricultural World Order: Regional Plant Protection Problems and International Phytopathology, 1878–1939'. *Agricultural History* 84, no. 1 (1 January 2010): 46–73.

Cavalcanti, Juliana Manzoni. 'Rudolf Kraus em busca do "ouro da ciência": A diversidade tropical e a elaboração de novas terapêuticas, 1913–1923'. *História, Ciências, Saúde-Manguinhos* 20, no. 1 (March 2013): 221–37.

Cayet, Thomas, Paul-André Rosental, and Marie Thébaud-Sorger. 'How International Organisations Compete: Occupational Safety and Health at the ILO, a Diplomacy of Expertise'. *Journal of Modern European History* 7, no. 2 (2009): 174–96.

Chakrabarti, Pratik. '"Living versus Dead": The Pasteurian Paradigm and Imperial Vaccine Research'. *Bulletin of the History of Medicine* 84, no. 3 (2010): 387–423.

Chakrabarty, Dipesh. *Provincializing Europe: Postcolonial Thought and Historical Difference*. Princeton, NJ: Princeton University Press, 2000.

Chlumecký, Johann von, ed. *Verhandlungen der internationalen Conferenz zur Erzielung eines gleichförmigen Vorgehens gegen die Rinderpest, gehalten vom 16. März bis 6. April 1872 in Wien*. Wien: Kaiserlich-Königliche Hof- und Staatsdruckerei, 1872.

Clavin, Patricia. *Securing the World Economy: The Reinvention of the League of Nations, 1920–1946*. Oxford: Oxford University Press, 2013.

Cliff, Andrew D., and Matthew Smallman-Raynor. *Oxford Textbook of Infectious Disease Control. A Geographical Analysis from Medieval Quarantine to Global Eradication*. Oxford: Oxford University Press, 2013.

Colijn, Hendrikus. 'The World Economic Conference of 1927'. *Annals of the American Academy of Political and Social Science* 134 (1 November 1927): 140–44.

Comité des Travaux Historiques et Scientifiques (Ecole Nationale des Chartes). 'Louis Camille Leblanc'. Annuaire Prosopographique: La France Savante. Comité des Travaux Historiques et Scientifiques (Ecole Nationale des Chartes), n.d. Accessed 18 March 2021. https://cths.fr/an/savant.php?id=4034.

Conrad, Sebastian. *What Is Global History?* Princeton: Princeton University Press, 2016.

Conrad, Sebastian, and Shalini Randeria. 'Globalgeschichte, Globalisierung, multiple Modernen: Zur Geschichtsschreibung der modernen Welt'. In *Globalgeschichte. Theorien, Ansätze, Themen*, edited by Sebastian Conrad, Andreas Eckert, and Ulrike Freitag, 7–49. Frankfurt am Main and New York: Campus, 2007.

Conrad, Sebastian, and Søren Tinning. 'Einleitung: Geteilte Geschichten – Europa in einer postkolonialen Welt'. In *Jenseits des Eurozentrismus. Postkoloniale Perspektiven in den Gechichts- und Kulturwissenschaften*, edited by Sebastian Conrad , Shalini Randeria and Regina Römhild, 32–70. 2nd ed. Frankfurt am Main: Campus, 2013.

Cooper, Anthony, and Søren Tinning. 'Introduction: Thinking Theoretically and Philosophically about the Concept of the Border'. In *Debating and Defining Borders: Philosophical and Theoretical Perspectives*, edited by Anthony Cooper and Søren Tinning, 1–13. Abingdon and New York: Routledge, 2019.

Critchell, James Troubridge, and Joseph Raymond. *A History of the Frozen Meat Trade*. London: Constable, 1912.

Cueto, Marcos. *The Value of Health: A History of the Pan American Health Organization*. Washington, D.C.: Pan American Health Organization, 2007.

Departmental Committee on Anthrax (Great Britain, Home Office). *Report of the Departmental Committee Appointed to Inquire as to Precautions for Preventing Danger of Infection from Anthrax in the Manipulation of Wool, Goat Hair, and Camel Hair*. London: H. M. Stationery Office, 1918.

Devoto, Luigi. 'The Milan Clinic for Occupational Diseases. Its Origin, Purposes and Activities'. In *Diseases of Occupation and Vocational Hygiene*, edited by George Martin Kober and William C. Hanson, 765–76. Philadelphia: P. Blakiston, 1916.

Di Fiore, Laura. 'The Production of Borders in Nineteenth-Century Europe: Between Institutional Boundaries and Transnational Practices of Space'. *European Review of History* 24, no. 1 (2016): 36–57.

Dieckerhoff, Wilhelm. *Geschichte der Rinderpest und ihrer Literatur. Beitrag zur Geschichte der vergleichenden Pathologie*. Berlin: T. C. F. Enslin, 1890.

Dorset, M., C. N. McBryde, and W. B. Niles. 'Remarks on "Hog" Flu'. *Journal of the American Veterinary Medical Association* 62 (1922): 162–71.

Driesch, Angela von den, and Peter Joris. *Geschichte der Tiermedizin. 5000 Jahre Tierheilkunde*. Stuttgart: Callwey, 2002.

Du Bois, W. E. B. 'Dantes Bellegarde: International Spokesman of Black Folk'. *The Crisis* 31 (April 1926): 295–96.

Edelmann, Richard H. 'Fleischbeschau'. In *Handbuch der Hygiene*, edited by Theodor Weyl, 3:475–546. Jena: Gustav Fischer, 1896.

Eijkman, Pieter H. *L'Internationalisme Médical*. Amsterdam: F. van Rossen, 1910.

———. *L'Internationalisme Scientifique: Sciences Pures et Lettres*. La Haye: W. P. van Stockum, 1911.

Engelken, Katharina, Dominik Hünniger, and Steffi Windelen, eds. *Beten, Impfen, Sammeln. Zur Viehseuchen- und Schädlingsbekämpfung in der Frühen Neuzeit*. Göttingen: Universitätsverlag Göttingen, 2007.

'Erwin Wehrle'. *Tierärztliche Rundschau* 42 (1936): 662.

Farewell, Vern, and Tony Johnson. 'Major Greenwood (1880–1949): A Biographical and Bibliographical Study'. *Statistics in Medicine* 35, no. 5 (28 February 2016): 645–70.

Ferrières, Madeleine. *Sacred Cow, Mad Cow: A History of Food Fears*. New York: Columbia University Press, 2006.

Fiennes, Richard N. *Zoonoses and the Origins and Ecology of Human Disease*. London and New York: Academic Press, 1978.

Fisher, John R. 'Cattle Plagues Past and Present: The Mystery of Mad Cow Disease'. *Journal of Contemporary History* 33, no. 2 (1998): 215.

Fitzgerald, Amy J. 'A Social History of the Slaughterhouse: From Inception to Contemporary Implications'. *Human Ecology Review* 17, no. 1 (2010): 58–69.

Fleming, George. *A Manual of Veterinary Sanitary Science and Police*. 2 vols. London: Chapman & Hall, 1875.

Flückiger, Gottlieb. 'Bisherige Bemühungen zur internationalen Bekämpfung tierischer Seuchen'. *Schweizer Archiv für Tierheilkunde* 96, no. 1 (1954): 12–29.

———. 'Zur Frage der Abschlachtung bei Maul- und Klauenseuche'. In *XVIII. Internationaler Landwirtschaftskongress Dresden, 6. bis 12. Juni 1939*, 3:44–52. Berlin: Reichsnährstand Verlags-Ges.m.b.H, 1939.

Food and Agriculture Organization of the United Nations. 'Reports of the Conference of FAO, Seventh Session, Rome November 23–December 11, 1953, Appendix: Agreement between FAO and OIE,' n.d. Accessed 19 March 2021. http://www.fao.org/docrep/x5576E/x5576e0h.htm#c.%20agreement%20between%20fao%20and%20oie.

Frevert, Ute. *Vertrauensfragen: Eine Obsession der Moderne*. München: Beck, 2013.

Friedrich-Loeffler-Institut (FLI). 'Friedrich-Loeffler-Institut: 1910–2010. FLI Riems: History and Development,' n.d. Accessed 19 March 2021. http://www.fli.bund.de/en/startseite/friedrich-loeffler-institut/historie/standort-insel-riems.html.

———. 'Friedrich-Loeffler-Institut: Historie – Standort Insel Riems,' n.d. Accessed 31 January 2021. https://www.fli.de/de/ueber-das-fli/historie/standort-insel-riems/.

Galli-Valerio, Bruno. *Zoonosi: Malattie Trasmissibili dagli Animali all'Uomo*. Milano: Ulrico Hoepli, 1894.

Gamgee, John. *The Cattle Plague. With Official Reports of the International Veterinary Congresses, Held in Hamburg, 1863, and in Vienna, 1865*. London: Hardwicke, 1866.

Gerabek, Werner E. 'Richard Pfeiffer'. In *Neue Deutsche Biographie*, 20:323. Berlin: Duncker & Humblot, 2001.

Gerber, Therese Steffen. 'Bürgi, Moritz'. *Historisches Lexikon der Schweiz*, n.d. Accessed 19 March 2021. http://www.hls-dhs-dss.ch/textes/d/D17456.php.

———. 'Flückiger, Gottlieb'. *Historisches Lexikon der Schweiz*, n.d. Accessed 19 March 2021. http://www.hls-dhs-dss.ch/textes/d/D17457.php.

Geyer, Martin H., and Johannes Paulmann, eds. *The Mechanics of Internationalism: Culture, Society, and Politics from the 1840s to the First World War*. Oxford and New York: Oxford University Press, 2001.

Giles-Vernick, Tamara, Susan Craddock, and Jennifer Lee Gunn, eds. *Influenza and Public Health: Learning from Past Pandemics*. London and Washington, D.C.: Earthscan, 2010.

Gilfoyle, Daniel. 'Anthrax in South Africa: Economics, Experiment and the Mass Vaccination of Animals, c. 1910–1945'. *Medical History* 50, no. 4 (2006): 465–90.

———. 'Veterinary Research and the African Rinderpest Epizootic: The Cape Colony, 1896–1898'. *Journal of Southern African Studies* 29, no. 1 (2003): 133–54.

Goodman, Neville M. *International Health Organizations and Their Work*. Edinburgh: Churchill Livingstone, 1971.

Gradmann, Christoph. *Krankheit im Labor: Robert Koch und die medizinische Bakteriologie*. Göttingen: Wallstein, 2005.

Greaves, H. R. G. *The League Committees and World Order: A Study of the Permanent Expert Committees of the League of Nations as an Instrument of International Government*. Oxford: Oxford University Press, 1931.

Greenhill, Robert. 'Shipping and the Refrigerated Meat Trade from the River Plate, 1900–1930'. *International Journal of Maritime History* 5, no. 1 (1992): 65–82.

Greenwood, Major. 'Dixième Rapport sur les Données Relatives au Traitement Antirabique fournies par les Instituts Pasteur'. *Bulletin de l'Organisation d'Hygiène* 12, no. 3 (1945): 319–84.

Greve, Bernard Anton. *Erfahrungen und Beobachtungen über die Krankheiten der Hausthiere im Vergleich mit den Krankheiten der Menschen*. 2 vols. Oldenburg: Schulze, 1818.

Gross Solomon, Susan, Lion Murard, and Patrick Zylberman, eds. *Shifting Boundaries of Public Health: Europe in the Twentieth Century*. Rochester, NY: University of Rochester Press, 2008.

Grut, A. 'The Work of the International Labour Organization in Occupational Health'. *British Journal of Industrial Medicine* 8, no. 4 (1951): 199–205.

Guerrini, Anita. *Experimenting with Humans and Animals. From Galen to Animal Rights*. Baltimore: Johns Hopkins University Press, 2003.

Haarmann, Daniela, and Kerstin Weich. 'Geschichte der Tiermedizin'. In *Tiere. Kulturwissenschaftliches Handbuch*, edited by Roland Borgards, 149–59. Stuttgart: Metzler, 2016.

Haas, Peter M. 'Introduction: Epistemic Communities and International Policy Coordination'. *International Organization* 46, no. 1 (1991): 1–35.

Haigh, Edward. *Remarks on the Subject of Rinderpest, Russian Plague, the Present Prevailing Disease amongst Cattle*. Manchester: No Publisher, 1865.

Hansen, Bert. 'America's First Medical Breakthrough: How Popular Excitement about a French Rabies Cure in 1885 Raised New Expectations for Medical Progress'. *American Historical Review* 103 (1998): 373–418.

Hardy, Anne. 'Actions Not Words. Thorvald Madsen, Denmark, and International Health, 1902–1939'. In *Of Medicine and Men. Biographies and Ideas in European Social Medicine between the World Wars*, edited by Iris Borowy and Anne Hardy, 127–43. Frankfurt am Main: Lang, 2008.

———. 'Animals, Disease, and Man: Making Connections'. *Perspectives in Biology and Medicine* 46, no. 2 (2003): 200–215.

———. 'Pioneers in the Victorian Provinces: Veterinarians, Public Health and the Urban Animal Economy'. *Urban History* 29, no. 3 (2002): 372–87.

———. 'Professional Advantage and Public Health: British Veterinarians and State Veterinary Services, 1865–1939'. *Twentieth Century British History* 14, no. 1 (2003): 1–23.

Harrison, Mark. 'A Global Perspective Reframing the History of Health, Medicine, and Disease'. *Bulletin of the History of Medicine* 89 (2015): 639–89.

———. *Contagion: How Commerce Has Spread Disease*. New Haven: Yale University Press, 2012.

———. 'Disease, Diplomacy and International Commerce: The Origins of International Sanitary Regulation in the Nineteenth Century'. *Journal of Global History* 1, no. 2 (2006): 197–217.

Häsler, Stephan. 'Die Veterinärabkommen des Völkerbundes von 1935: Eine Vision des Veterinärraumes Europa'. *Schweizer Archiv für Tierheilkunde* 160, no. 1 (2018): 37–42.

Hering, Eduard, and Wilhelm Probstmayr, eds. *Amtlicher Bericht über die internationale thierärztliche Versammlung zu Hamburg am 14.–18. Juli 1863*. Stuttgart: Ebner & Seubert, 1864.

Herren, Madeleine. 'Fascist Internationalism'. In *Internationalisms. A Twentieth-Century History*, edited by Glenda Sluga and Patricia Clavin, 191–212. Cambridge: Cambridge University Press, 2017.

———. *Hintertüren zur Macht: Internationalismus und modernisierungsorientierte Aussenpolitik in Belgien, der Schweiz und den USA, 1865–1914*. München: Oldenbourg, 2000.

———. 'International Organizations, 1865–1945'. In *The Oxford Handbook of International Organizations*, edited by Jacob Katz Cogan, Ian Hurd, and Ian Johnstone, 91–112. Oxford: Oxford University Press, 2016.

———. *Internationale Organisationen seit 1865. Eine Globalgeschichte der internationalen Ordnung*. Darmstadt: Wissenschaftliche Buchgesellschaft, 2009.

———. 'Introduction: Towards a Global History of International Organization'. In *Networking the International System: Global Histories of International Organizations*, edited by Madeleine Herren, 1–12. Heidelberg: Springer, 2014.

———. '"Outwardly…an Innocuous Conference Authority": National Socialism and the Logistics of International Information Management'. *German History* 20, no. 1 (2002): 67–92.

Herren, Madeleine, and Sacha Zala. *Netzwerk Aussenpolitik. Internationale Kongresse und Organisationen als Instrumente der schweizerischen Aussenpolitik 1914–1950*. Zürich: Chronos, 2002.

Herring, Ann, and Alan C. Swedlund, eds. *Plagues and Epidemics: Infected Spaces Past and Present*. Oxford and New York: Berg, 2010.

Hetherington, H. J. W. *International Labour Legislation*. London: Methuen, 1920.
Hobson, Asher. *The International Institute of Agriculture: An Historical and Critical Analysis of Its Organization, Activities and Policies of Administration*. Berkeley: University of California Press, 1931.
Hörning, B., and R. Fankhauser. 'Vor 100 Jahren: 6. Internationaler Tierärztlicher Kongress, September 1895 in Bern'. *Schweizer Archiv fur Tierheilkunde* 138, no. 1 (1996): 8–11.
Howell, Philip, and Hilda Kean, eds. *The Routledge Companion to Animal-Human History*. London and New York: Routledge, 2018.
Hoy, Suellen, and Walter Nugent. 'Public Health or Protectionism? The German-American Pork War, 1880–1891'. *Bulletin of the History of Medicine* 63 (1989): 198–224.
Huber, Valeska. '"The Unification of the Globe by Disease"? The International Sanitary Conferences on Cholera, 1851–1894'. *Historical Journal* 49, no. 2 (2006): 453–76.
Hubscher, Ronald. *Les Maîtres des Bêtes: les Vétérinaires dans la Société Française, XVIIIe-XXe Siècle*. Paris: Odile Jacob, 1999.
Hünniger, Dominik. *Die Viehseuche von 1744–52. Deutungen und Herrschaftspraxis in Krisenzeiten*. Neumünster: Wachholtz Verlag, 2011.
Hüntelmann, Axel. 'Geschichte des Tierversuchs'. In *Tiere. Kulturwissenschaftliches Handbuch*, edited by Roland Borgards, 160–73. Stuttgart: Metzler, 2016.
Hunter, Pamela. *Veterinary Medicine: A Guide to Historical Sources*. Aldershot: Ashgate, 2004.
Hutyra, Ferenc. 'Bericht Entwurf einer internationalen Veterinär-Konvention'. In *Siebenter Internationaler Tierärztlicher Kongress: Baden-Baden, 7.–12. August 1899*, 1:5–32. Baden-Baden: Hofdruckerei Ernst Kölblin, 1899.
———. 'Internationale Tierseuchen-Polizei'. In *VIe Congrès International de Médecine Vétérinaire, tenu à Berne, du 16 au 21 Septembre 1895: Rapports et Compte-Rendu*, edited by Emile Noyer, 173–200. Bern: Stämpfli, 1896.
Hutyra, Ferenc, and József Marek. *Die Orientalische Rinderpest. Mit besonderer Berücksichtigung der klinischen und anatomischen Merkmale und der Differentialdiagnose*. Jena: Fischer, 1916.
———. *Spezielle Pathologie und Therapie der Haustiere*. Jena: Fischer, 1905.
'Imperial Bureaux'. *The British Medical Journal*, no. 3546 (22 December 1928), 1148–1149.
Insenhöfer, Swantje. 'Dr. Friedrich Weber. Reichstierärzteführer von 1934 bis 1945'. PhD Thesis, Tierärztliche Hochschule Hannover, 2008.
Institut Pasteur. 'Institut Pasteur – Archives: Dossiers Biographiques. Albert Calmette (1863–1933),' n.d. Accessed 1 April 2021. https://webext.pasteur.fr/archives/cal0.html.
———. 'Institut Pasteur – Archives: Dossiers Biographiques. Emile Roux (1853–1933)', n.d. Accessed 1 June 2020. https://webext.pasteur.fr/archives/rou0.html.
———. 'Institut Pasteur: Gaston Ramon – Pro Humanitate,'n.d.-c. Accessed 1 April 2021. https://www.pasteur.fr/fr/institut-pasteur/notre-histoire/gaston-ramon-pro-humanitate.
———. 'Institut Pasteur: Repères Chronologiques. Auguste-Charles Marie (1864–1935)', n.d. Accessed 1 April 2021. https://web.archive.org/web/20110823091919/http://www.pasteur.fr/infosci/archives/mca0.html.

———. 'Institut Pasteur: Repères Chronologiques. Henri Pottevin (1865–1928),' n.d. Accessed 1 April 2021. https://webext.pasteur.fr/archives/pot0.html.

———. 'Institut Pasteur: Repères Chronologiques. Paul Remlinger (1871–1964),' n.d. Accessed 19 March 2020. https://web.archive.org/web/20101221105424/http://www.pasteur.fr/infosci/archives/rem0.html.

International Institute of Agriculture/Institut International d'Agriculture. *Le Commerce International de la Viande*. Rome: Institut International d'Agriculture, 1936.

———. *Quelques Aspects de l'Activité de l'Institut International d'Agriculture (1905–1940)*. Rome: Institut International d'Agriculture, 1941.

———, ed. *The Prevention of Anthrax Infection among Live Stock. Results of an International Enquiry*. Rome: Printing Office of the Chamber of Deputies, 1925.

International Labour Office, ed. *Advisory Committee on Anthrax, London, December 1922. Report of the Committee*. Geneva: International Labour Office, 1923.

———. *Disinfection of Wool Infected with Anthrax Spores: Questionnaire*. Geneva: International Labour Office, 1920.

———. *International Labour Conference, 7th Session, Geneva 1925: Report on Workmen's Compensation*. Geneva: International Labour Office, 1925.

———. 'Prévention de l'Infection Charbonneuse parmi les Troupeaux'. *ILO Official Bulletin* 8 (1923): 104.

International Labour Organisation, ed. *ILO Yearbook*. Geneva: International Labour Office, 1930.

———. 'C018: ILO Workmen's Compensation (Occupational Diseases) Convention, 1925 (No. 18)'. NORMLEX: Information System on International Labour Standards, n.d. Accessed 19 March 2021. http://www.ilo.org/dyn/normlex/en/f?p=1000:12100:0::NO::P12100_INSTRUMENT_ID:312163.

International Labour Organization/Organisation Internationale du Travail. 'Première Session de La Commission Consultative d'Hygiène Industrielle'. *ILO Bulletin Officiel* 8 (Juillet–Décembre 1923): 134–35.

Jackson, Alan C., and William H. Wunner. *Rabies*. Amsterdam and London: Elsevier/Academic, 2007.

Jahr, Christoph, and Stefan Kaufmann. 'Den Krieg führen: Organisation, Gewalt, Technik'. In *Erster Weltkrieg: Kulturwissenschaftliches Handbuch*, edited by Niels Werber, Stefan Kaufmann, and Lars Koch, 164–231. Stuttgart: Metzler, 2014.

Jones, Bruce Vivash. 'Unfortunate Timing…'. *Veterinary Record* 174, no. 25 (2014): 627–29.

Jones, David, and Stefan Helmreich. 'A History of Herd Immunity'. *The Lancet* 396, no. 10254 (September 2020): 810–11.

Jones, Susan D. *Death in a Small Package. A Short History of Anthrax*. Baltimore: Johns Hopkins University Press, 2010.

Jong, A. de. 'Préface'. In *Travaux Du IXe Congrès International de Médecine Vétérinaire à La Haye, 13–19 Septembre 1909*, 3:1–4. Leiden: Brill, 1910.

Kaiser, Wolfram, and J. W. Schot. *Making Europe: Experts, Cartels, and International Organizations*. Basingstoke: Palgrave Macmillan, 2014.

Kastner, Justin. 'Food and Agriculture Security: An Historical Illustration of Contemporary Challenges'. In *Food and Agriculture Security: A Historical, Multidisciplinary Approach*, 5–32. Santa Barbara: Praeger, 2011.

———. 'Managing Human and Animal-Health Threats: Additional Lessons from the 19th-Century Trading World'. In *Food and Agriculture Security: A Historical, Multidisciplinary Approach*, 105–27. ABC-CLIO, 2011.

Kete, Kathleen. 'La Rage and the Bourgeoisie: The Cultural Context of Rabies in the French Nineteenth Century'. *Representations* 22, no. 1 (1988): 89–107.

Knab, Cornelia. 'Infectious Rats and Dangerous Cows: Transnational Perspectives on Animal Diseases in the First Half of the Twentieth Century'. *Contemporary European History* 20, no. 3 (2011): 281–306.

Kohlmeyer, Frederick William. 'The Movement toward International Cooperation in Food and Agriculture: Background of the FAO of the UN'. PhD Thesis, University of Minnesota, 1954.

Kohlrausch, Martin. 'Technologische Innovation und transnationale Netzwerke. Europa zwischen den Weltkriegen/Technological Innovation and Transnational Networks: Europe between the Wars'. *Journal of Modern European History* 6, no. 2 (2008): 181–95.

Kohlrausch, Martin, Katrin Steffen, and Stefan Wiederkehr. 'Expert Cultures in Central Eastern Europe'. In *Expert Cultures in Central Eastern Europe. The Internationalization of Knowledge and the Transformation of Nation States since World War I*, edited by Martin Kohlrausch, Katrin Steffen, and Stefan Wiederkehr, 9–30. Osnabrück: Fibre, 2010.

Koolmees, Peter A. 'Veterinary Inspection and Food Hygiene in the Twentieth Century'. In *Food, Science, Policy, and Regulation in the Twentieth Century: International and Comparative Perspectives*, edited by Jim Phillips and David F. Smith, 53–68. London and New York: Routledge, 2000.

Koren, Nathan. *Jewish Physicians: A Biographical Index*. John Wiley and Sons, 1973.

Kraus, Rudolf, Franz Gerlach, and Fritz Schweinburg. *Lyssa bei Mensch und Tier*. Berlin und Wien: Urban & Schwarzenberg, 1926.

Langner, Hans. 'Tagung zur Gründung einer zwischenstaatlichen Arbeitsgemeinschaft auf veterinärem Gebiet'. *Deutsches Tierärzteblatt. Amtsblatt der Reichstierärztekammer* 10, no. 15/16 (1943): 85–90.

———. 'Tagung zur Gründung einer zwischenstaatlichen Arbeitsgemeinschaft auf veterinärem Gebiet (Fortsetzung)'. *Deutsches Tierärzteblatt. Amtsblatt der Reichstierärztekammer* 10, no. 17/18 (1943): 95–98.

Laqua, Daniel, ed. *Internationalism Reconfigured: Transnational Ideas and Movements between the World Wars*. London and New York: Tauris, 2011.

Law, James. 'Report of the International Veterinary Congress of Brussels'. Edited by United States Bureau of Animal Industry. *Annual Report of the Bureau of Animal Industry*, 321–70, 1885.

League of Nations. 'Veterinary Questions, 1. General Report of the Sub-Committee of Experts on Veterinary Questions, November 4, 1930. Document C. 633.M.252.1930.II,' n.d. Accessed 1 April 2021. https://biblio-archive.unog.ch/Dateien/CouncilMSD/C-633-M-252-1930-II_EN.pdf.

League of Nations Economic and Financial Section: Monetary and Economic Conference. *Report and Proceedings of the World Economic Conference held at Geneva, May 4th to 23rd, 1927*. 2 vols. Geneva: League of Nations, 1927.

League of Nations Information Section. *Health Organisation*. Geneva: League of Nations, 1931.

———. *The Health Organisation of the League of Nations*. Geneva: League of Nations, 1923.

League of Nations, and International Labour Office, eds. *Conférence Internationale du Travail/International Labour Conference, 3rd Session Geneva 1921*. Geneva: International Labour Office, 1921.

League of Nations and International Labour Office eds. *Conférence Internationale du Travail/International Labour Conference, 6th Session Geneva 1924*. Geneva: International Labour Office, 1924.

———. *International Labor Conference. First Annual Meeting, October 29, 1919–November 29, 1919*. Washington, D.C.: Government Printing Office, 1920.

Leblanc, Camille. 'Rapport: Service Sanitaire International'. In *Ve Congrès International de Médecine Vétérinaire, tenu à Paris du 2 au 8 Septembre 1889: Recueil de Rapports*, edited by Saturnin Arloing, 145–53. Paris: Asselin et Houzeau, 1889.

Leblanc, Camille, and Paul Cagny, eds. *Bulletin Trimestriel International des Epizooties*, 1er et 2e trimestres 1883. Paris: Recueil de Médecine Vétérinaire, 1883.

Leclainche, Emmanuel. 'Die derzeitige Verbreitung der Maul- und Klauenseuche und die Möglichkeiten ihrer Prophylaxe'. In *XVIII. Internationaler Landwirtschaftskongress Dresden, 6. bis 12. Juni 1939*, 3, Sektion VI: Tierzucht und Tierhaltung:20–25. Berlin: Reichsnährstand Verlags-Ges.m.b.H, 1939.

———. *Histoire de la Médecine Vétérinaire*. Toulouse: Office du Livre, 1936.

———. *Histoire Illustrée de la Médecine Vétérinaire*. 2 vols. Paris: A. Michel, 1955.

———. 'La Situation Sanitaire de l'Europe en 1946'. *OIE Bulletin* 26 (1946): 15–23.

———. 'La "Standardisation" des Bulletins Sanitaires'. *OIE Bulletin* 1 (Juillet 1927–Juin 1928), 577–83.

———. 'L'Epizootie Aphteuse de 1937 à 1946'. *OIE Bulletin* 26 (1946): 24–33.

———. 'Un Concept d'une Police Sanitaire Moderne'. *OIE Bulletin* 6 (Mai 1932–Avril 1933), 173–174.

Lee, Paula Young, ed. *Meat, Modernity, and the Rise of the Slaughterhouse*. Durham, NH: University of New Hampshire Press, 2008.

Leymann, Hermann. *Anthrax in the Tannery Industry*. Edited by International Labour Office. Geneva: International Labour Office, 1923.

Lhoste, Albert. *Les Conventions Vétérinaires Internationales*. Paris: Vigot Frères, 1936.

Lichte, Jan Ulrich. 'Die Forschung auf der Insel Riems von 1933 bis 1945 unter besonderer Berücksichtigung der NS-Zwangsarbeiter'. PhD Thesis, Ernst-Moritz-Arndt-Universität Greifswald, 2011.

Lombard, M., P. P. Pastoret, and A. M. Moulin. 'A Brief History of Vaccines and Vaccination'. *Revue Scientifique et Technique (OIE)* 26, no. 1 (2007): 29–48.

London School of Economics, Library. 'Record: Greaves, Harold Richard Goring,' n.d. Accessed 19 March 2021. http://archives.lse.ac.uk/Record.aspx?src=CalmView.Catalog&id=COLL+MISC+0462.

LONSEA. 'International Institute of Refrigeration/Institut International du Froid,' n.d. Accessed 12 March 2021. http://www.lonsea.de/pub/org/257.

———. 'Permanent Committee for International Veterinary Congresses,' n.d. Accessed 19 March 2020. http://www.lonsea.de/pub/org/107.

———. 'Searching the Globe through the Lenses of the League of Nations,' n.d. Accessed 30 June 2020. http://www.lonsea.de/.

Lubinski, Herbert. 'Zur Ätiologie der nach Schutzimpfung gegen Lyssa vorkommenden Lähmungen'. *Klinische Wochenschrift* 5, no. 31 (1 July 1926): 1419–21.

Ludwig, Peter. *Die Arbeiten des Internationalen Tierseuchenamtes in Paris seit seiner Gründung im Jahre 1921 bis zum Ausbruch des europäischen Krieges im September 1939*. Bern: J. Werder, 1941.

Machado, Manuel A. *Aftosa: A Historical Survey of Foot-and-Mouth Disease and Inter-American Relations*. Albany: State University of New York Press, 1969.

Magallanes, Nelson. 'Ciento Veinte Años de Aftosa en Uruguay'. *Veterinaria* 29, no. 120 (Dicembre 1993): 4–18.

Mammerickx, Marc. 'La Peste Bovine, Jules Bordet et le Centre Sérumigène de Cureghem'. *Annales de Médecine Vétérinaire* 147 (2003): 197–205.

Mantovani, Adriano. 'Notes on the Development of the Concept of Zoonoses'. *Historia Medicinae Veterinaria* 26, no. 2 (2001): 41–53.

Mantovani, Adriano, and R. Zanetti. 'Giovanni Maria Lancisi: De Bovilla Peste and Stamping Out'. *Historia Medicinae Veterinariae* 18, no. 4 (1993): 97–110.

Marie, Auguste C., Paul Remlinger, and Henri Vallée, eds. 'Conférence Internationale de la Rage'. *Annales de l'Institut Pasteur* 42, no. 12, Supplément (Dec. 1928).

Mazumdar, Pauline M. H. '"In the Silence of the Laboratory": The League of Nations Standardizes Syphilis Tests'. *Social History of Medicine* 16, no. 3 (2003): 437–59.

McKendrick, Anderson Gray. 'Cinquième Revue Analytique des Rapports des Instituts Pasteur sur les Résultats de la Vaccination Antirabique'. *Bulletin Trimestriel de l'Organisation d'Hygiène* 3 (1934): 647–79.

———. 'Huitième Revue Analytique des Rapports des Instituts Pasteur sur les Résultats de la Vaccination Antirabique'. *Bulletin de l'Organisation d'Hygiène* 7 (1938): 5–45.

———. *Première Revue Analytique des Rapports des Instituts Pasteur sur les Résultats de la Vaccination Antirabique*. Genève: Société des Nations, 1930.

McVety, Amanda Kay. *The Rinderpest Campaigns: A Virus, Its Vaccines, and Global Development in the Twentieth Century*. Cambridge: Cambridge University Press, 2018.

McVey, Scott, and Jishu Shi. 'Vaccines in Veterinary Medicine: A Brief Review of History and Technology'. *Veterinary Clinics of North America: Small Animal Practice* 40, no. 3 (2010): 381–92.

Melikoglu, B., R. T. Bagasaç Gül, and T. Özkul. 'Paul Ambroise Remlinger: A Pasteurien in Turkey and His Studies on Rabies'. *Revue de Médecine Vétérinaire* 160, no. 7 (2009): 374–77.

Meyer, Karl Friedrich. *The Zoonoses in their Relation to Rural Health*. Berkeley: University of California Press, 1955.

Mezzadra, Sandro, and Brett Neilson. 'Foreword'. In *Debating and Defining Borders: Philosophical and Theoretical Perspectives*, edited by Anthony Cooper and Søren Tinning, XVII–XXV. Abingdon and New York: Routledge, 2019.

Miescher, Giorgio. *Namibia's Red Line: The History of a Veterinary and Settlement Border*. New York: Palgrave Macmillan, 2012.

Mishra, Saurabh. *Beastly Encounters of the Raj: Livelihoods, Livestock and Veterinary Health in India, 1790–1920*. Manchester: Manchester University Press, 2016.

———. 'Beasts, Murrains, and the British Raj: Reassessing Colonial Medicine in India from the Veterinary Perspective, 1860–1900'. *Bulletin of the History of Medicine*, 2011, 587–619.

Moore, John. *Army Veterinary Service in War*. London: Brown, 1921.

Morens, David M., Jeffrey K. Taubenberger, Gregory K. Folkers, and Anthony S. Fauci. 'An Historical Antecedent of Modern Guidelines for Community Pandemic Influenza Mitigation'. *Public Health Reports* 124 (February 2009): 22–24.

Moulin, Anne Marie. 'The Pasteur Institute's International Network: Scientific Innovation and French Tropisms'. In *Transnational Intellectual Networks. Forms of Academic Knowledge and the Search for Cultural Identities*, edited by Christophe Charle, Jürgen Schriewer, and Peter Wagner, 135–64. Frankfurt am Main and New York: Campus, 2004.

Nathusius, Hermann von, and Anton Krocker. *Landwirthschaftliches Thier-Album in Photographien: Die Internationale Landwirthschaftliche Ausstellung in Hamburg, 1863*. Berlin: Wiegand und Hempel, 1866.

Nicolas, Eugène, and Paul Rinjard. *La Production du Sérum contre la Peste Bovine au Centre Sérumigène de Cureghem-lez-Bruxelles*. Toulouse: Imprimerie Ouvrière, 1922.

Nieradzik, Lukas. 'Geschichte der Nutztiere'. In *Tiere. Kulturwissenschaftliches Handbuch*, edited by Roland Borgards, 121–29. Stuttgart: Metzler, 2016.

Noyer, Emile, ed. *VIe Congrès International de Médecine Vétérinaire, tenu à Berne, du 16 au 21 Septembre 1895: Rapports et Compte-Rendu*. Bern: Stämpfli, 1896.

O'Hern, E. M. 'Alice Evans and the Brucellosis Story'. *Annali Sclavo: Rivista di Microbiologia e di Immunologia* 19, no. 1 (1977): 12–19.

Olmstead, Alan L. 'The First Line of Defense: Inventing the Infrastructure to Combat Animal Diseases'. *The Journal of Economic History* 69, no. 2 (2009): 327–57.

Olmstead, Alan L., and Paul W. Rhode. 'An Impossible Undertaking: The Eradication of Bovine Tuberculosis in the United States'. *Journal of Economic History* 64, no. 3 (2004): 734–72.

———. *Arresting Contagion: Science, Policy, and Conflicts over Animal Disease Control*. Cambridge, MA.: Harvard University Press, 2015.

———. 'The "Tuberculous Cattle Trust": Disease Contagion in an Era of Regulatory Uncertainty'. *The Journal of Economic History* 64, no. 4 (2004): 929–63.

'One Medicine? Investigating Human and Animal Disease,' n.d. Accessed 5 March 2021. http://www.kcl.ac.uk/artshums/depts/history/research/proj/onemedicine.aspx.

Orland, Barbara. 'Handeln in Zeiten der Ungewißheit. Tuberkulose, Milch und Tierseuchenbekämpfung im 19. und 20. Jahrhundert'. *Mittelungen des Internationalen Arbeitskreises für Kulturforschung des Essens* 8 (May 2001): 13–24.

Osterhammel, Jürgen. 'Global History and Historical Sociology'. In *The Prospect of Global History*, edited by James Belich, John Darwin, Margret Frenz, and Chris Wickham, 23–43. Oxford and New York: Oxford University Press, 2016.

———. 'Globalizations'. In *The Oxford Handbook of World History*, edited by Jerry H. Bentley, 89–104. Oxford and New York: Oxford University Press, 2011.

———. *The Transformation of the World: A Global History of the Nineteenth Century*. Princeton: Princeton University Press, 2014.

Österreich-Ungarn, K. K. 'Ackerbau-Ministerium'. In *Bericht über die Thätigkeit des K. K. Ackerbau-Ministeriums in der Zeit vom 1. Jänner 1869 bis 30. Juni 1874. Erster Theil*. Wien: Faesy und Frick, 1874.

Ostertag, Robert von. *Bibliographie der Fleischbeschau. Zugleich Ergänzung zum Handbuch der Fleischbeschau*. Stuttgart: F. Enke, 1905.

———. *Handbook of Meat Inspection*. 2nd ed. London: Baillière & Co., 1905.

Otter, Christopher. 'Civilizing Slaughter: The Development of the British Public Abattoir, 1850–1910'. *Food and History* 3, no. 2 (2008): 29–51.

Özkul, Türel, and R. Tamay Başagaç Gül. 'La Collaboration de Maurice Nicolle et Adil Mustafa. La Découverte de l'Agent de la Peste Bovine'. *Revue de Médecine Vétérinaire* 159, no. 4 (2008): 243–46.

Pacyga, Dominic A. *Chicago: Slaughterhouse to the World*. In *Meat, Modernity, and the Rise of the Slaughterhouse,* edited by Paula Young Lee, 153–66. Durham, NH: University of New Hampshire Press, 2008.

Pan American Health Organization. 'Veterinary Public Health – PAHO/WHO,' n.d. Accessed 1 April 2021. http://www.paho.org/en/topics/veterinary-public-health.

Pattison, Iain. *John McFadyean: A Great British Veterinarian*. London and New York: J. A. Allen, 1981.

Pemberton, Neil, and Michael Worboys. *Mad Dogs and Englishmen: Rabies in Britain, 1830–2000*. Houndmills, Basingstoke: Palgrave Macmillan, 2007.

Perren, Richard. *Taste, Trade and Technology: The Development of the International Meat Industry since 1840*. Aldershot: Ashgate, 2006.

'Personalien – Geheimer Oberregierungsrat Dr. August Lydtin'. *Schweizer Archiv für Tierheilkunde* 59 (1917): 642–43.

Philips, Howard. 'The Re-Appearing Shadow of 1918: Trends in the Historiography of the 1918–19 Influenza Pandemic'. *Canadian Bulletin of Medical History = Bulletin Canadien d'Histoire de la Médecine* 21, no. 1 (2004): 121–34.

Phoofolo, Pule. 'Face to Face with Famine: The BaSotho and the Rinderpest, 1897–1899'. *Journal of Southern African Studies* 29, no. 2 (1 June 2003): 503–27.

Podgorny, Irina. 'A Horse-Cloth for Uganda, or How an Account by a Transhumant Veterinary Connects Histories, Animal Diseases and Continents'. *History and Technology* 34, no. 1 (2018): 71–78.

Pottevin, Henri. 'Rapport sur la Pandémie Grippale de 1918–1919 présenté au Comité Permanent de l'Office International d'Hygiène Publique'. *Bulletin de l'Office International d'Hygiène Publique* 13 (1921): 125–81.

Probstmayr, Wilhelm. *Etymologisches Wörterbuch der Veterinär-Medicin und ihrer Hilfswissenschaften*. München: Grubert, 1863.

Proust, Adrien. *La Défense de l'Europe contre la Peste et la Conférence de Venise de 1897*. Paris: Masson, 1897.

———. *La Défense de l'Europe contre le Choléra*. Paris: Masson, 1892.

Pschorr, Wilhelm. 'Karl Dammann'. In *Neue Deutsche Biographie*, 3:500–501. Berlin: Duncker & Humblot, 1957.

Rátz, István, ed. *Bericht über den VIII. internationalen tierärztlichen Kongress in Budapest 1905*. 3 vols. Budapest: Pátria, 1906.

'Raymond Gautier, M.D.'. *The British Medical Journal* 1, no. 5027 (1957): 1127.

Reinisch, Jessica. 'Internationalism in Relief: The Birth (and Death) of UNRRA'. *Past and Present* 210, Supplement no. 6 (2011): 258–89.

———. 'Introduction: Relief in the Aftermath of War'. *Journal of Contemporary History* 43, no. 3 (2008): 371–404.

Reis, Ursula. *Tierseuchen und internationaler Handel. Veterinärexperten im Völkerbund 1927–1935*. Universität Zürich: Lizentiatsarbeit, unpublished, 2005.

République Française, Ministère de l'Agriculture, ed. *Conférence Internationale pour l'Etude des Epizooties. Paris (25–28 Mai 1921). Compte Rendu Sténographique*. Paris: Imprimerie Nationale, 1921.

République Française, Ministère de l'Agriculture et du Commerce, ed. *Documents sur la Peste Bovine*. Paris: Imprimerie Nationale, 1872.

Reuter, M. 'Tierseuchen und sporadische Tierkrankheiten im Kriege. Deren Auftreten und Bekämpfung bis zum Stande vom 1. Mai 1916'. In *Ergebnisse der Hygiene, Bakteriologie, Immunitätsforschung und experimentellen Therapie*, edited by W. Weichardt, 2:668–747. Berlin: Springer, 1917.

Reynal, Jean. *Traité de la Police Sanitaire des Animaux Domestiques*. Paris: Asselin, 1873.

Ricard, Joseph. *Les Origines de la Première Conférence Internationale des Epizooties, Fièvre Aphteuse, Peste Bovine*. Paris: de Chaix, 1921.

Ritvo, Harriet. 'On the Animal Turn'. *Daedalus* 136, no. 4 (2007): 118–22.

Rizzo Naudi, John. *Brucellosis – The Malta Experience: A Celebration, 1905–2005*. San Gwann (Malta): Publishers Enterprise Group (PEG), 2005.

Rodogno, Davide, eds. *Shaping the Transnational Sphere. Experts, Networks and Issues from the 1840s to the 1930s*. New York and Oxford: Berghahn Books, 2015.

Rodogno, Davide, Bernhard Struck, and Jakob Vogel. 'Introduction'. In *Shaping the Transnational Sphere. Experts, Networks and Issues from the 1840s to the 1930s*, 1–20. New York and Oxford: Berghahn Books, 2015.

Röll, Moritz Friedrich, and Leopold Forster. *Amtlicher Bericht über den zweiten internationalen Congress von Thierärzten zu Wien am 21. – 17. August 1865*. Wien: Wilh. Braumüller, 1865.

Roman, Agnes. 'UNRRA in China'. *Far Eastern Survey* 14, no. 22 (7 November 1945): 320–24.

Rosenau, Milton J. 'The Number of Bacteria in Milk and the Value of Bacterial Counts'. In *Milk and Its Relation to the Public Health*, edited by US National Institutes of Health, 420–45. Washington, D.C. Government Printing Office, 1908.

Ruzicka, Jan, and Vincent Charles Keating. 'Going Global: Trust Research and International Relations'. *Journal of Trust Research* 5, no. 1 (2015): 8–26.

Samanta, Samiparna. 'Dealing with Disease: Epizootics, Veterinarians and Public Health in Colonial Bengal, 1850–1920'. In *Medicine and Colonialism: Historical Perspectives in India and South Africa*, edited by Poonam Bala, 61–74. London and New York: Routledge, 2015.

Sarasin, Philipp. *Anthrax: Bioterror as Fact and Fantasy*. Cambridge, MA.: Harvard University Press, 2006.

———. 'Die Visualisierung des Feindes. Über metaphorische Technologien der frühen Bakteriologie'. *Geschichte und Gesellschaft* 30, no. 2 (2004): 250–76.

Sawyer, Wilbur A. 'Achievements of UNRRA as an International Health Organization'. *American Journal of Public Health and the Nation's Health* 37, no. 1 (January 1947): 41–58.

Scherer, Sarah Brian. 'Potterat, Denys'. *Historisches Lexikon der Schweiz*, n.d. Accessed 19 March 2020. http://mobile.hls-dhs-dss.ch/m.php?article=D17449.php.

Schern, Kurt. 'Bericht aus London'. *Berliner Tierärztliche Wochenschrift* 30, no. 38 (1914): 677–78.

Schmaltz, Reinhold. 'Dem Jubiläumskongreß!' *Berliner Tierärztliche Wochenschrift* 30, no. 32 (1914): 565–66.

———. *Entwicklungsgeschichte des tierärztlichen Berufes und Standes in Deutschland*. Berlin: Schoetz, 1936.

———. 'Jubiläumskongreß in Waffen'. *Berliner Tierärztliche Wochenschrift* 30, no. 33 (1914): 597–98.

———. 'Zur tierärztlichen Internationale'. *Berliner Tierärztliche Wochenschrift* 33, no. 17 (1917): 202.

Schultz-Cherry, Stacey, Christopher W. Olsen, and Bernard C. Easterday. 'History of Swine Influenza'. In *Swine Influenza*, edited by Jürgen A. Richt and Richard J. Webby, 21–27. Berlin, Heidelberg: Springer, 2013.

Schutte, C. E., ed. *Comptes-Rendus du Congrès International tenu à Prétoria du 2 au 13 août 1897 pour l'Etude de la Peste Bovine et des Moyens de Combattre cette Epizootie dans l'Afrique Méridionale.* Pretoria: Staatsdrukkerij van de Zuid-Afrikaansche Republiek, 1898.

Schwabe, Calvin W. *Veterinary Medicine and Human Health.* 2nd ed. Baltimore: Williams & Wilkins, 1984.

Sclavo, Achille. 'Anthrax'. In *Occupation and Health. Encyclopaedia of Hygiene, Pathology, and Social Welfare,* No. 36, edited by International Labour Office: ILO Correspondence Committee on Industrial Hygiene, 1:2–11. Geneva: International Labour Office, 1930.

Sealey, Anne. 'Globalizing the 1926 International Sanitary Convention'. *Journal of Global History* 6, no. 3 (2011): 431–55.

Siebenter Internationaler Tierärztlicher Kongress: Baden-Baden, 7.–12. August 1899. 2 vols. Baden-Baden: Hofdruckerei Ernst Kölblin, 1899.

Sinclair, Upton. *The Jungle.* New York: Doubleday, Page & Co., 1906.

Sluga, Glenda. *Internationalism in the Age of Nationalism.* Philadelphia: University of Pennsylvania Press, 2013.

Sluga, Glenda, and Patricia Clavin, eds. *Internationalisms: A Twentieth-Century History.* Cambridge: Cambridge University Press, 2017.

Smith, David F., and Jim Phillips, eds. *Food, Science, Policy, and Regulation in the Twentieth Century: International and Comparative Perspectives.* London and New York: Routledge, 2000.

Société des Nations, ed. 'Rapport sur l'Oeuvre de l'Organisation d'Hygiène de Juin 1938 à Avril 1939 et sur son Programme de Travail 1939'. *Bulletin de l'Organisation d'Hygiène* 8 (1939): 1–62.

Société des Nations/League of Nations. *Répertoire des Organisations Internationales/Handbook of International Organisations.* Geneva: League of Nations, 1921.

Sommer, Louise. 'Die Vorgeschichte der Weltwirtschaftskonferenz (Genf 1927)'. *Weltwirtschaftliches Archiv* 28 (1928): 340–418.

Spiekermann, Uwe. 'Dangerous Meat? German-American Quarrels over Pork and Beef'. *Bulletin of the GHI Washington* 46 (Spring 2010): 93–110.

Spinage, Clive A. *Cattle Plague: A History.* Dordrecht and London: Kluwer, 2003.

Staples, Amy L. S. *The Birth of Development. How the World Bank, Food and Agriculture Organization, and World Health Organization Changed the World, 1945–1965.* Kent, Ohio: Kent State University Press, 2006.

Steele, James H. 'A Bookshelf on Veterinary Public Health'. *American Journal of Public Health* 63, no. 4 (1973): 291–311.

———. 'Zoonoses 1879–1983: An Update of James Law's Report on Diseases of Animals'. *Veterinary Heritage* 23, no. 1 (2000): 14–20.

Stockman, Stewart, ed. *Tenth International Veterinary Congress.* 3 vols. London: J. Bale Sons & Danielsson, 1915.

Stordy, Robert J. 'Belgium and East Africa. Danger to a Great Cattle Area. Letters to the Editor'. *The Times*, 24 May 1919.

Stümke, Hans, ed. *Bibliographie der internationalen medizinischen Kongresse und Verbände*. Bibliographie der internationalen Kongresse und Verbände 1. Berlin: O. Harrassowitz, 1939.

Sunseri, Thaddeus. 'The African Rinderpest Panzootic, 1888–1897'. Oxford Research Encyclopedia of African History, 26 April 2018. https://oxfordre.com/africanhistory/view/10.1093/acrefore/9780190277734.001.0001/acrefore-9780190277734-e-375.

———. 'The Entangled History of Sadoka (Rinderpest) and Veterinary Science in Tanzania and the Wider World, 1891–1901'. *Bulletin of the History of Medicine* 89, no. 1 (2015): 95–121.

Swabe, Joanna. *Animals, Disease, and Human Society. Human-Animal Relations and the Rise of Veterinary Medicine*. London and New York: Routledge, 1999.

Taithe, Bertrand. *Defeated Flesh: Medicine, Welfare, and Warfare in the Making of Modern France*. Manchester: Manchester University Press, 1999.

Taylor, William P., Peter L. Roeder, and Mark M. Rweyemamu. 'History of Vaccines and Vaccination'. In *Rinderpest and Peste des Petits Ruminants. Virus Plagues of Large and Small Ruminants*, edited by William P. Taylor, Thomas Barrett, and Paul-Pierre Pastoret, 222–46. London and Burlington, MA: Elsevier/Academic, 2006.

Teissier, Marie. 'A Brief History of the OIE'. *OIE Bulletin*, no. 1 (2007): 6–18.

'The Risk of Rinderpest. Belgian Measures in East Africa'. *The Times*, 28 May 1919.

The Avalon Project. Documents in Law, History and Diplomacy. 'The Versailles Treaty: Part XIII, Labour,' n.d. Accessed 19 March 2021. http://avalon.law.yale.edu/imt/partxiii.asp.

'The War Day by Day'. *The Times*, 1 August 1914.

'The War Day by Day'. *The Times*, 4 August 1914.

Théodoridès, Jean. 'Bref Historique de la Fièvre de Malte'. *Histoire des Sciences Médicales* 30, no. 1 (1996): 87–90.

Thomassen, M. H. J. P. 'Rapport: Service Sanitaire International'. In *Ve Congrès International de Médecine Vétérinaire, tenu à Paris du 2 au 8 Septembre 1889: Recueil de Rapports*, edited by Saturnin Arloing, 87–143. Paris: Asselin et Houzeau, 1889.

Thompson, Ruth D'Arcy. *The Remarkable Gamgees: A Story of Achievement*. Edinburgh: Ramsay Head Press, 1974.

Todd, Charles. *Description of the Serum Institute (Abbassia) and of the Preparation of Cattle Plague Serum*. Cairo: Govt. Press, 1912.

Tollardo, Elisabetta. *Fascist Italy and the League of Nations, 1922–1935*. London: Palgrave Macmillan, 2016.

Toma, Bernard. 'A Propos de l'Origine des Mots "épizootique" et "épizootie"'. *Bulletin de la Société Française d'Histoire de la Médecine et des Sciences Vétérinaires* 7 (2007): 26–49.

Travaux du IXe Congrès International de Médecine Vétérinaire à La Haye, 13–19 Septembre 1909. 3 vols. Leiden: Brill, 1910.

Trischler, Helmuth, and Martin Kohlrausch. *Building Europe on Expertise: Innovators, Organizers, Networkers*. Basingstoke: Palgrave Macmillan, 2014.

Turnbull, Peter C. B., and Sean Shadony. 'Anthrax from 5000 BC to AD 2010'. In *Bacillus Anthracis and Anthrax*, edited by Nicholas H. Bergman, 1–15. Hoboken, NJ: Wiley-Blackwell, 2011.

United Nations Treaty Collection. 'International Convention Concerning the Export and Import of Animal Products (Other than Meat, Meat Preparations, Fresh Animal Products, Milk and Milk Products). Geneva, 20 February 1935'. United

Nations Treaty Collection, n.d. Accessed 19 February 2021. https://treaties.un.org/Pages/LONViewDetails.aspx?src=LON&id=570&chapter=30&clang=_en

———. 'International Convention Concerning the Transit of Animals, Meat and Other Products of Animal Origin. Geneva, 20 February 1935'. United Nations Treaty Collection, n.d. Accessed 19 February 2021. https://treaties.un.org/Pages/LONViewDetails.aspx?src=LON&id=569&chapter=30&clang=_en

———. 'International Convention for the Campaign against Contagious Diseases of Animals. Geneva, 20 February 1935'. United Nations Treaty Collection, n.d. Accessed 19 February 2021. https://treaties.un.org/Pages/LONViewDetails.aspx?src=LON&id=568&chapter=30&clang=_en

———. 'Pacte de la Société des Nations/Covenant of the League of Nations'. United Nations Treaty Collection: League of Nations Treaty Series, n.d. Accessed 1 March 2021. https://treaties.un.org/doc/source/covenant.pdf.

———. 'Protocol Concerning the Office International d'Hygiène Publique. New York, 22 July 1946'. United Nations Treaty Collection, n.d. Accessed 19 March 2021. http://treaties.un.org/doc/Treaties/1947/10/19471020%2006-23%20AM/Ch_IX_02p.pdf.

Universität Greifswald. 'Die Universität Greifswald im Nationalsozialismus. Die besonderen Beziehungen zur Insel Riems', n.d. Accessed 1 May 2021. https://ns-zeit.uni-greifswald.de/themengebiete/kriegseinsatz/kriegswichtige-forschungen-und-profilwandel/die-besonderen-beziehungen-zur-insel-riems/.

Universiteit Utrecht. 'Catalogus Professorum: Professoren van de Universiteit Utrecht sinds 1636: Prof. Dr. H. C. L. E. Berger', n.d. Accessed 30 May 2021. http://profs.library.uu.nl/index.php/profrec/getprofdata/149/2/30/0.

———. 'Catalogus Professorum: Professoren van de Universiteit Utrecht sinds 1636: Prof. Dr. L. de Blieck' n.d. Accessed 19 March 2020. http://profs.library.uu.nl/index.php/profrec/getprofdata/192/1/9/0.

Vallat, François. *Les Bœufs Malades de la Peste. La Peste Bovine en France et en Europe, XVIIIe–XIXe Siècle*. Rennes: Presses Universitaires de Rennes, 2009.

Veyssière, Jean-Baptiste-Hippolyte. *Des Maladies Transmissibles des Animaux à l'Homme*. Paris: Libraires de l'École de Médecine, 1852.

Virchow, Rudolf. 'Intoxicationen, Zoonosen und Syphilis'. In *Handbuch der speciellen Pathologie und Therapie*, edited by Frieder Alexander Simon, Carl P. Falck, and Rudolf Virchow, 2.1:337–420. Erlangen: F. Enke, 1855.

Waddington, Keir. *The Bovine Scourge: Meat, Tuberculosis and Public Health, 1850–1914*. Woodbridge: Boydell Press, 2006.

———. 'To Stamp out "so Terrible a Malady": Bovine Tuberculosis and Tuberculin Testing in Britain 1890–1939'. *Medical History* 48, no. 1 (2004): 29–48.

Waldmann, Otto. 'La Vaccination contre la Fièvre Aphteuse'. *OIE Bulletin* 21 (1942): 133–49.

———. 'Sur l'Aspect Futur de la Lutte contre la Fièvre Aphteuse'. *OIE Bulletin* 21 (1942): 331–57.

Wall, Rosemary. *Bacteria in Britain, 1880–1939*. Abingdon and New York: Routledge, 2016.

Walters, F. P. *A History of the League of Nations*. London and New York: Oxford University Press, 1952.

Wedekind, Klemens. *Impfe und herrsche: Veterinärmedizinisches Wissen und Herrschaft im kolonialen Namibia 1887–1929*. Göttingen: Vandenhoeck & Ruprecht, 2021.

Weech, Friedrich von, ed. 'Christian Josef Fuchs'. In *Badische Biographien, 1. Lieferung*, 270–71. Heidelberg: Bassermann, 1875.

Wehenkel, Jean, ed. *Compte-Rendu du IVe Congrès International de Médecine Vétérinaire, Bruxelles 1883*. Bruxelles: Brogniez et Vande Weghe, 1884.

Wehrle, Erwin. 'Geschichte der internationalen tierärztlichen Kongresse'. *Berliner Tierärztliche Wochenschrift* 30, no. 32 (1914): 567–81.

Weindling, Paul, ed. *International Health Organisations and Movements, 1918–1939*. Cambridge and New York: Cambridge University Press, 1995a.

———. 'Social Medicine at the League of Nations Health Organisation and the International Labour Office Compared'. In *International Health Organisations and Movements, 1918–1939*, edited by Paul Weindling, 134ff. Cambridge: Cambridge University Press, 1995b.

Wenzlhuemer, Roland. *Connecting the Nineteenth-Century World: The Telegraph and Globalization*. Cambridge and New York: Cambridge University Press, 2015.

———. 'Connections in Global History'. *Comparativ* 29, no. 2 (2019): 106–21.

Wilkinson, Lise. *Animals and Disease. An Introduction to the History of Comparative Medicine*. Cambridge and New York: Cambridge University Press, 1992.

———. 'Glanders: Medicine and Veterinary Medicine in Common Pursuit of a Contagious Disease'. *Medical History* 25, no. 4 (1981): 363–84.

———. 'Rinderpest and Mainstream Infectious Disease Concepts in the Eighteenth Century'. *Medical History* 28, no. 2 (1984): 129–50.

Woodbridge, George. *UNRRA. The History of the United Nations Relief and Rehabilitation Administration*. 3 vols. New York: Columbia University Press, 1950.

Woods, Abigail. *A Manufactured Plague: The History of Foot-and-Mouth Disease in Britain*. London: Earthscan, 2004.

———. 'From Practical Men to Scientific Experts: British Veterinary Surgeons and the Development of Government Scientific Expertise, c. 1878–1919'. *History of Science* 51, no. 4 (2013): 457–80.

———. 'Is Prevention Better than Cure? The Rise and Fall of Veterinary Preventive Medicine, c.1950–1980'. *Social History of Medicine* 26, no. 1 (2013): 113–31.

———. 'The Construction of an Animal Plague: Foot-and-Mouth Disease in Nineteenth-Century Britain'. *Social History of Medicine* 17, no. 1 (2004): 23–39.

Woods, Abigail, Michael Bresalier, Angela Cassidy, and Rachel Mason Dentinger. *Animals and the Shaping of Modern Medicine: One Health and Its Histories*. London and New York: Palgrave Macmillan, 2018.

Worboys, Michael. 'Germ Theories of Disease and British Veterinary Medicine, 1860–1890'. *Medical History* 35 (1991): 308–27.

———. *Spreading Germs: Disease Theories and Medical Practice in Britain, 1865–1900*. Cambridge and New York: Cambridge University Press, 2000.

———. 'Was There a Bacteriological Revolution in Late Nineteenth-Century Medicine?' *Studies in History and Philosophy of Biological and Biomedical Sciences* 38, no. 1 (2007): 20–42.

World Health Organization. *Joint WHO/FAO Expert Committee on Zoonoses. Second Report, Stockholm, 11–16 August 1958*. World Health Organization Technical Report Series 169. Geneva: World Health Organization, 1959.

———. 'Joint WHO/FAO Expert Committee on Zoonoses. Second Report, "Rabies"', n.d. Accessed 19 March 2021. https://www.who.int/health-topics/rabies#tab=tab_1.

World Organisation for Animal Health (OIE). 'Animal Diseases: Portal (OIE List)', n.d. Accessed 30 May 2021. https://www.oie.int/en/what-we-do/animal-health-and-welfare/animal-diseases/.

———. 'Appendix to the International Agreement: Organic Statutes of the Office International Des Epizooties,', n.d. Accessed 1 April 2021. https://www.oie.int/about-us/key-texts/basic-texts/organic-statutes/.

———. 'COVID-19', n.d. Accessed 30 May 2021. https://www.oie.int/en/what-we-offer/emergency-and-resilience/covid-19/.

———. 'FAO/OIE Global Conference on Foot and Mouth Disease Control', n.d. Accessed 19 March 2021. http://www.oie.int/eng/A_FMD2012/background.html.

———. 'History', n.d. Accessed December 28 2021. https://www.oie.int/en/who-we-are/mission/history/.

———. 'International Agreement for the Creation of an Office International des Epizooties in Paris, done in Paris, 25 January 1924', n.d. Accessed 15 May 2021. https://www.oie.int/about-us/key-texts/basic-texts/international-agreement-for-the-creation-of-an-office-international-des-epizooties/.

———. 'No More Deaths from Rabies,' n.d. Accessed 1 April 2021. http://www.oie.int/infographic/rabies/.

———. 'One Health', n.d. Accessed 28 May 2021. http://www.oie.int/en/for-the-media/onehealth/.

———. 'Rabies Portal', n.d. Accessed 25 April 2021. http://www.oie.int/animal-health-in-the-world/rabies-portal/.

World Organisation for Animal Health (OIE), and Food and Agriculture Organization of the United Nations (FAO). 'Final Report of the Joint Committee on Global Rinderpest Eradication, May 2011', n.d. Accessed 15 May 2021. http://www.oie.int/doc/ged/D10943.PDF.

World Trade Organization. 'WTO Legal Texts – The General Agreement on Tariffs and Trade (GATT 1947)', n.d. Accessed 1 April 2021. https://www.wto.org/english/docs_e/legal_e/gatt47_02_e.htm#articleXX.

Wrenn, John Edwin, and Frank M. Surface. *International Trade in Meats and Animal Fats*. Washington, DC: U.S. Govt. Printing Office, 1925.

XVIII. Internationaler Landwirtschaftskongress Dresden, 6. bis 12. Juni 1939, Sektion VI: Tierzucht und Tierhaltung, Spezialberichte. Thema 1: Die neuesten Erfahrungen bei der Bekämpfung der Maul- und Klauenseuche. Berlin: Reichsnährstand Verlags-GMBH, 1939.

Zaidi, Waqar. 'Liberal Internationalist Approaches to Science and Technology in Interwar Britain and the United States'. In *Internationalism Reconfigured. Transnational Ideas and Movements between the World Wars*, edited by Daniel Laqua, 17–43. London and New York: Tauris, 2011.

Zangger, Rudolf, ed. *Amtlicher Bericht über den dritten internationalen Congress von Thierärzten zu Zürich am 2.–7. September 1867*. Zürich: Schabelitz, 1869.

Zundel, (Auguste). 'Rapport: Organisation de Service Vétérinaire'. In *Compte-Rendu du IVe Congrès International de Médecine Vétérinaire, Bruxelles 1883*, edited by Jean Wehenkel, 99–144. Bruxelles: Brogniez et Vande Weghe, 1884.

Zürcher, Christoph. 'Schenk, Carl'. *Historisches Lexikon der Schweiz*, n.d. Accessed 17 October 2020. https://hls-dhs-dss.ch/articles/004636/2011-07-12/.

Zylberman, Patrick. 'Civilizing the State. Borders, Weak States and International Health in Modern Europe'. In *Medicine at the Border. Disease, Globalization and Security, 1850 to the Present*, edited by Alison Bashford, 21–40. Houndmills, Basingstoke: Palgrave Macmillan, 2006.

———. 'Making Food Safety an Issue: Internationalized Food Politics and French Public Health from the 1870s to the Present'. *Medical History* 48, no. 1 (2004): 1–28.

Index